LOCAL FOOD NETWORKS AND ACTIVISM IN THE HEARTLAND

THOMAS R. SADLER, EDITOR
HEATHER MCILVAINE-NEWSAD, EDITOR
BILL KNOX, EDITOR

LOCAL FOOD NETWORKS AND ACTIVISM IN THE HEARTLAND

THOMAS R. SADLER, EDITOR
HEATHER MCILVAINE-NEWSAD, EDITOR
BILL KNOX, EDITOR

COMMON GROUND

First published in 2013 in Champaign, Illinois, USA
by Common Ground Publishing LLC
as part of The Food Studies Book Series

Library of Congress Cataloging-in-Publication Data

Local food networks and activism in the heartland / Thomas R. Sadler, Heather McIlvaine-Newsad, Bill Knox, editors.
p. cm.
Includes bibliographical references.
ISBN 978-1-61229-196-3 (pbk : alk. paper) -- ISBN 978-1-61229-197-0 (pdf)
1. Local foods--United States. 2. Food supply--United States. 3. Community-supported agriculture--United States. I. Sadler, Thomas R. II. McIlvaine-Newsad, Heather, 1967- III. Knox, Bill, 1951-

HD9005.L62 2013
338.1'977--dc23

2013010448

Table of Contents

Acknowledgements

The editors would like to thank Common Ground Publishing and especially Ian Nelk and Brian Kornell for their helpful comments and suggestions. We appreciate working with this publishing company. We would also like to thank the many reviewers who read each chapter, provided useful feedback, and enhanced the quality of the writing.

List of Contributors

Rachel Armstrong, Postdoctoral Research Associate, University of Illinois Urbana-Champaign

Leslie Cooperband, Soil Scientist, Co-Owner and Cheesemaker, Prairie Fruits Farm and Creamery, Champaign, Illinois

John Curtis, Barefoot Gardens CSA, Macomb, Illinois

A. Brian Endres, Associate Professor of Agricultural Law, Department of Agricultural and Consumer Economics, University of Illinois Urbana-Champaign

Joel B. Gruver, Assistant Professor of Agriculture, School of Agriculture, Western Illinois University

Molly E. Hicks, Doctoral Candidate, Department of Environmental Studies, Antioch University

Frederick Iutzi, Program Manager, Value Added Sustainable Development Center, Illinois Institute for Rural Affairs, Western Illinois University

Wesley Jarrell, Professor Emeritus of Sustainable Agriculture and Natural Resources, Department of Natural Resources and Environmental Sciences, University of Illinois Urbana-Champaign and Co-Owner, Prairie Fruits Farm and Creamery, Champaign, Illinois

Bill Knox, Professor of English, Department of English and Journalism, Western Illinois University

Jamie Lane, Registered Nurse, Intensive Care Unit, Cottage Hospital, Galesburg, Illinois

Christopher D. Merrett, Director, Illinois Institute for Rural Affairs, Western Illinois University

Heather McIlvaine-Newsad, Professor of Anthropology, Department of Sociology and Anthropology, Western Illinois University

Erin Orwig, Faculty Assistant, Illinois Institute for Rural Affairs, Western Illinois University

Betsy Perabo, Associate Professor of Religious Studies, Department of Philosophy and Religious Studies, Western Illinois University

Rob Porter, Assistant Professor of Recreation, Park and Tourism Administration, Department of Recreation, Park and Tourism Administration, Western Illinois University

Brian K. Powell, Associate Professor of Philosophy, Department of Philosophy and Religious Studies, Western Illinois University

Thomas R. Sadler, Associate Professor of Economics and Director, Center for Economic Education, Department of Economics and Decision Sciences, Western Illinois University

Lisa Schlessinger, Postdoctoral Research Associate, University of Illinois Urbana-Champaign

Sarah Shike, Faculty Assistant, Illinois Institute for Rural Affairs and Lecturer, Department of Management and Marketing, Western Illinois University

Holly A. Stovall, Assistant Professor of Women's Studies, Department of Women's Studies, Western Illinois University

Preface

This book is for all people interested in local food networks and activism. In its chapters, voices from different walks of life—local food growers, academics interested in sustainable food production and consumption, and local food activists—write about contemporary challenges concerning the local food movement. Real-world examples such as the birth of a community garden, the success of a local food initiatives group, the importance of a seed bank, situating the local food movement in the context of the legal system, the sale of locally-sourced produce, the challenges of social justice and morality, and many others fill the book's pages.

The idea for the book grew out of a class on sustainable food networks at Western Illinois University during the spring semester of 2011. Taught by economics faculty member Thomas R. Sadler, this honors course brought together a bright group of students with a professor who wanted to explore the growing importance and many nuances of local food networks. The class adopted an interdisciplinary framework with a series of presentations on the benefits and costs of local food networks and the importance of local food activism.

The course was a huge success. Not only did weekly discussions address fascinating topics such as the history of industrial agriculture, national and local food chains, morality and food production, organic farming, food and feminist theory, community development, activism, public policy, food politics, international trade, food security, food and energy, the importance of cooking, human health, cultural knowledge and gardening, but the course included several dynamic guest speakers.

The guest speakers were experts on local food networks and activism. During the presentations, it became clear that we, as an academic and activist community in the heartland, collectively possess a strong and interdisciplinary knowledge base concerning local food. To take four examples from the course, Dr. Joel Gruver discussed agriculture in the context of the organic Allison farm that he operates north of Macomb, Illinois. Dr. Holly Stovall discussed cooking and feminist theory. Dr. Brian Powell considered the moral foundations of the local food movement. Dr. Heather McIlvaine-Newsad addressed food, anthropology and the birth of a community garden.

The guest speakers from the sustainable food course along with course's professor are the co-authors of this book. Additional co-authors were invited to contribute chapters after making fascinating presentations in October, 2012 on local food production, activism and the legal system at the 2nd Annual Food Studies Conference, in Champaign, Illinois, organized by Common Ground.

A common theme throughout the book is that a local food network improves the balance of a given community and economy. For example, local food providers who want to sell directly to consumers take advantage of market opportunities, including farmers markets, community-supported agriculture (CSA), and farm stands. This opportunity allows farmers and gardeners selling produce to keep a larger percentage of the revenue from their sales. As another example, many individuals want to provide food for their families that contributes to healthy diets, not diminishes or compromises health. Local and organic foods provide such an opportunity. A final example entails community activism that builds the institutions and knowledge base necessary to sustain the local food movement, including producers, markets and cooking. This activism brings like-minded individuals together to discuss growing techniques, local food markets and legal and social impediments, but also to share good, wholesome meals.

The setting of the book is the "heartland," which generally means the agricultural middle of the United States, but specifically refers to the small part of the world in which the co-authors reside: central Illinois. As many chapters explain, this part of the country exhibits the characteristics of many agricultural areas: rural landscapes of monoculture and small towns, but a growing awareness of the importance of locally-sourced food.

Many years ago, the writer and farmer Wendell Berry (1981) explained how industrial agriculture methods "destroy, first, the health of the soil and finally the health of human communities." While this description holds true for many acres of land and several struggling towns in the region, many others, including Macomb and Champaign, both in central Illinois, have many locally-sourced food options. The writers in this book demonstrate how the growth of local food markets in this region improves the quality of the soil and the health of human communities, thus "solving for pattern" and causing a "ramifying series of solutions" to specific economic, environmental and social problems (Berry, 1981).

However, even though the co-authors of this book reside in the heartland, the themes are relevant to all people interested in local food production, consumption and activism. Families that garden together or discuss healthy food options often appreciate the availability of locally-sourced food. Families and friends that take time to prepare savory meals and enjoy them at a leisurely pace may appreciate having the knowledge of where their food comes from.

These are ideas that Ben Hewitt (2009) addresses in his thoughtful book, *The Town That Food Saved*. Even though Hewitt's setting, Hardwick, Vermont, possesses a history and culture of small farming—the opposite of the large-scale farms that exist in central Illinois—the goals that Hewitt writes of that include building local food networks, adding jobs in local food production, and linking the growth of a local food economy to the resurgence of communities also constitute our goals.

The knowledge and wisdom that accompany a vibrant local food network also enhance the appreciation of the connection between food and health issues. For women and men, spending time in the kitchen preparing thoughtful meals for family and friends is an act of resilience and a lifestyle choice. In many instances, a third place—neither home nor work—at a community garden or a community-supported agricultural program brings people together from different socio-economic backgrounds to share conversation, community news, and the challenges of growing vegetables.

Stories of the local food movement in this book discuss many important questions: Do the decisions we make concerning food production and consumption enhance or diminish environmental quality? With respect to our food choices, how should we treat animals? Do our farming practices and consumption decisions contribute to community development? If it is important to develop a local food economy, how should we build the necessary institutions? In what circumstances is cooking a feminist act? What are the legal implications of local food networks? How does a diet that consists largely of locally-sourced food contribute to human health? By addressing these and many other questions, the book adds to the growing literature on local food networks and activism, including *The Town That Food Saved* (Hewitt, 2009), *The Seasons on Henry's Farm* (Brockman, 2009), and, of course, *The Omnivore's Dilemma* (Pollan, 2006).

While providing unique research and writing concerning the local food movement in the heartland, the book also addresses important points of many influential food writers. Writing about the extraordinary science of addictive junk food, Michael Moss (2013) discusses how, with processed food, food engineers alter many variables with the intent of finding the perfect version of the product. The result is the addictive behavior of consumers to processed food. Mark Bittman (2012) writes that public policy has been lacking in addressing many leading health challenge of the 21st century: not finding cures for killers such as diabetes and coronary artery diseases. Raj Patel (2010), in a narrative on the global food system, argues that malnutrition and food insecurity exist on a large scale because of the exploitive nature of the industrial food system. But the stories and research on local food networks in this book demonstrate the nutritional and community values associated with the consumption of sustainable food. Healthier and local food options not only provide a method of addressing the addictive nature and unhealthy outcomes of processed food, but they stop the exploitation of both farmers and consumers. When we use the term "local food," we don't have a particular radius in mind, such as the 100-mile diet. With the term local food, we mean a connection exists between food producer and consumer. Often, as in the case of the community-supported agriculture in Macomb, Illinois, the gardener, John Curtis, knows the shareholders' families, because they come on a weekly basis to his beautiful garden to harvest, kids and all.

Organization and Content

The book is organized according to four themes:

1. Local food and sustainability
2. Local food production
3. Local food and health
4. Local food and activism

Each area of focus raises certain questions. In the first section on local food and sustainability, the chapters address the challenges of establishing sustainable food networks, the goals of a local food network, the development of local food markets, and why a sense of place impacts the context in which local food networks develop. The second section addresses the challenges of local food production, including direct-to-consumer outlets, growth of a local food business, the development of seed banks, legal and institutional challenges, and the growing importance of community supported agriculture. The third section on local food and health considers why cooking constitutes a crucial aspect of the local food movement and why the consumption of locally-sourced food may lead to greater human health. The final section on local food and activism links the local food movement to religious, moral and socially-just outcomes.

Section One: Local Food and Sustainability

The Introductory chapter argues that the state of Illinois sends most of its food dollars to other states. Keeping a larger percentage of this spending local would lead to a fundamental shift in the way we think about food production and consumption. This chapter discusses the fact that people respond to incentives. Locally-sourced food that is both available and healthy is impacting the way people think about eating. The consequences of producing and consuming more local food entail the strengthening of a community's natural, cultural, human and social capital. A focus on the concept of networks provides a useful way to think about the development of local food economies. Networks change and evolve over time.

In Chapter 2 on "Sustainable Local Food: Turning the Soil, Turning the Tide," Bill Knox writes of his own background in sustainable and local food production. Influenced by parents who spent part of their childhoods around gardens and neighbors who kept large and productive gardens, he was impressed with the ideas of self-reliance, nurturing and generosity. Invited to live on a dairy farm in his twenties, the author learned how to plant, cultivate and harvest, enjoying the output of his labor. These impressions led to a lifetime of backyard planting, involvement in community gardens, and simpler tools and healthier behavior. As Knox explains, neighborhood gardening enhances human health and energy independence. The willingness to serve in the vanguard of the local food movement represents progressive behavior.

Chapter 3, "Where Does 'Local' Fit Under the Sustainability Tent? A Conversation with Dr. Joel Gruver," offers a definition of sustainable agriculture that includes social, economic and environmental concerns. An academic and expert on organic agriculture and local food networks, Dr. Gruver's concept of sustainability is important, because many chapters address ecological concerns such as the health of the soil, economic considerations such as the sale of locally-sourced produce, and social concerns such as the availability of fresh food for low-income households. After Dr. Gruver discusses his background and the organic farm he operates, he addresses the relationship between organic and sustainable, organic production and economies of scale, why local food networks matter, and the local food movement in central Illinois.

In Chapter 4 on "More than Just Food: Wisdom, Vitality and the Local Food Movement," Thomas R. Sadler argues that three goals of a strong local food network—growth of the demand and supply sides of the market, the creation of economic opportunity in the form of jobs, and the improvement of environmental and social conditions—require an understanding of the structure of a local food marketplace. Such a marketplace, when thriving and fully developed, is based on the values of wisdom, vitality, collective sustainability and economics. This marketplace includes direct-to-consumer outlets such as farmers markets, roadside stands, farm stores, community supported agriculture, food co-ops and food trucks. By using a framework of evaluation that includes variables such as econetwork services, cultural services, externality elimination and community development, this chapter concludes that the local food network in central Illinois is currently small but beginning to offer a diverse set of healthy food options.

In Chapter 5 on "A Sense of Place," John Curtis emphasizes that the landscape in central Illinois—thousands of acres of corn and soybeans in a land of monoculture—has harmed small towns and created an environment for industrial scale animal confinements. Historically, the agricultural environment was not like this. Small family farms with garden plots and a few animals dotted the landscape. A natural and circular method of farming included solar power, animals fertilizing future crops, oats and hay on-site, and time and care for food preparation. If the future of agriculture is to honor health as well as profit, Curtis argues that we must grow locally-sourced food for people in the region. A decentralized food network practiced on many scales is important for the health of the people, land and communities.

Section Two: Local Food Production

In Chapter 6 on "Prairie Fruits Farm and Creamery, Champaign, Illinois," Wesley Jarrell and Leslie Cooperband, in an original and informative narrative, discuss the influences and origins of their local food farm and business. Citing influences such as a personal search for sustainability, the writers Sir Albert Howard, Lady Eve Balfour, J.I. Rodale and Jerry Goldstein, and academic careers in soil science, Jarrell and Cooperband discuss how they started a fruit farm and then added a goat dairy, cheeserie and sheep milk. They now supply local and regional markets with their output, provide on-site dinners, and offer frequent tours. They also provide educational opportunities and an important model for farmers and other entrepreneurs wanting to start businesses in the local food market.

In Chapter 7, "Legal and Institutional Issues in Local Food Production: Lessons from Illinois," A. Bryan Endres, Lisa Schlessinger and Rachel Armstrong examine the "intuitional biases" in the system of food production that stem from the legislative and regulatory process. Focusing on local food production, the authors argue that a web of federal and state laws regulate or exempt local food producers from the larger system of operational and food safety regulations. As the direct farm business model grows, so does its vulnerability to the legal environment. However, despite institutional and legal barriers, the future of the local food movement is bright. Farmer and attorney education along with affordable legal service will move the entrepreneurial direct market farmer to a more stable future.

In Chapter 8, "Seeds and Sustainability: Why Keeping Seeds Local is an Act of Resistance and Resilience," Molly E. Hicks considers the important role of seeds within local and sustainable agri-food networks. Until the emergence of commercial seed networks, many farmers and backyard gardeners saved seeds. But the current reliance on seeds from commercial providers does not allow the planter to know how the seeds are grown or when they are harvested, processed or stored. In contrast, the process of saving seeds for the next growing season assures for future sustenance and increases individual and community food security. In seed exchanges where seeds are given or traded between communities, farmers and gardeners may exchange essential agricultural knowledge and tips regarding planting techniques. Seed saving and exchanges also promote crop diversity and enhance the local nature of the growing process.

In Chapter 9, "Trends in Local Food Production: The Growing Importance and Impacts of Community Supported Agriculture in the United States," Christopher D. Merrett, Frederick Iutzi, Erin Orwig, Sarah Shike and Heather

McIlvaine-Newsad examine the community supported agriculture (CSA) model of production, where families or individuals purchase a share from a local farmer or gardener and receive produce for the entire growing season. Arguing that the CSA model is important in expanding local food networks, this chapter reports the results of two national surveys on CSA farmers from 2002 and 2009. The survey results demonstrate that, instead of the economic bottom line, CSA operators are motivated by the ideas of environmental sustainability, fresh food and community development.

Section Three: Local Food and Health

The links between eating local foods and good health are often overshadowed by equally important questions about community building, the environment, agricultural challenges, and the political obstacles to creating local foods systems. Holly Stovall and Jamie Lane fill in some of the links about health, the former with a theoretical analysis and the latter with a model of how to eat locally for health.

In Chapter 10, "Let the Flavors Marry: A Gender-Integrated Conceptualization of Nourishment for All People," Holly A. Stovall, in her analysis based on Evelyn Fox Keller's *Gender and Science*, develops the thesis that separation from the source of our food has resulted in a nutritional crisis rooted in beliefs that nutrition "science" is a substitute for engaging in an intimate "relationship" with food and its origins and processing. In part of a discourse that privileges masculinity over femininity, the scientific lab and the consumer of food are conceptually separated from and made superior to the kitchen and the food itself. Stovall says that in order to resolve diet-caused diseased epidemics, consumers of food for nourishment must ask intimate and contextualized questions about what they eat. Scientific generalizations like "eat a Mediterranean diet," or "eat low-fat," encourage consumers of food for nourishment to rely on the institutional science, rather than on their own relationship with and knowledge about the food that goes on their tables.

In Chapter 11, "Eating Locally, Staying Healthy: A Nurse's Perspective," Jamie Lane and Holly A. Stovall, write about the fact that, Lane, an avid vegetable gardener and health-care practitioner, engages in an intimate relationship with her food from soil to table. As Lane describes, she knows the soil she works and cares for, the bees that pollinate the food she grows, the particular varieties of her fruits and vegetables, and even the "weeds," which, importantly, she often harvests for their flavor and nutritional value. Lane tells how this gardening life, inspired by her grandmother who ate almost exclusively homegrown food, is a recipe for a long life lived disease-free.

Section Four: Local Food Activism

In Chapter 12, "The Birth of the Macomb Community Garden," Heather McIlvaine-Newsad and Rob Porter write that their motivation in establishing a community garden is not only planting and harvesting, but the establishment of a socially just way of providing healthy food options for people of all socio-economic backgrounds. Concerned about the globalization of the food system, the authors created a third place—neither home nor work—where community members may gather and network. The authors argue that, the more diverse and robust the community garden, the more resilient our lives will be. In fact, the community garden enhances food democracy by helping to determine local food policies and practices. Sharing seeds and the harvest, the community garden members reclaim unused but fertile urban space, enjoy the opportunity to grow produce in a community environment, and address the problem of food insecurity that some community members face.

In Chapter 13, "It's Not About Food Miles: The Moral Foundations of the Local Food Movement," Brian K. Powell adds an important viewpoint to the book. The decisions we make concerning how we feed ourselves impact the lives of other people, non-human animals and the planet. An example is the meat produced in factory farms, which leads to health-related problems for people, suffering for non-human animals, and environmental degradation in the areas immediately adjacent to concentrated animal feeding operations. The local food movement, and its important moral component, entails knowledge of the process of food production. This knowledge leads to an understanding of where food comes from, how it is grown and distributed, how animals are treated, and the impacts on the environment. With this knowledge, the moral heart of the local food movement is the willingness of individuals to take responsibility for their food choices.

In Chapter 14, "What People are for: Agrarianism and Religion in the Heartland," Betsy Perabo argues that, for some people, rural areas are ideally suited for both religion and farming. Addressing the connection between these two, a "cottage industry" surrounding the local food movement and the concept of sustainability have emerged in U.S. non-native religious traditions and in religious studies. Ecological concerns have come from a more diverse set of academics who are seeking to situate it in different biblical and theological frameworks. The crucial question in addressing the relationship between religion and farming is: what are people for? By reviewing the literature on the role of religion in sustainability and the local food movement, Perabo addresses this question and demonstrates why U.S. non-native religious traditions teach that we must, if we want to be closer to a higher power, attend to agrarian concerns.

In the concluding Chapter 15, Heather McIlvaine-Newsad argues that the U.S. food system is polarized. In the industrial food industry, an unparalleled concentration of power exists with a few corporate giants, but at the same time local food networks continue to grow. Reflecting on her upbringing, which included time with her grandmother cooking delicious meals with locally-sourced ingredients (from their farm), McIlvaine-Newsad writes that as a child she knew where her food came from. For all of us, the knowledge of where our food comes from means we have the responsibility for protecting the health of the environment and ourselves. By being active in the local food movement and consuming locally-sourced food, we are helping to make sustainable and moral food choices.

References

Berry, Wendell. 1981. Solving for Pattern. In *The Gift of Good Land: Further Essays Cultural and Agricultural*. Berkeley: Counterpoint.

Bittman, Mark. 2012. Dietary Seat Belts. *The New York Times*, 18 December.

Brockman, Terra. 2009. *The Seasons on Henry's Farm*, Chicago: Surrey Books.
Hewitt, Ben. 2009. *The Town That Food Saved.* New York: Rodale.
Moss, Michael. 2013. The Extraordinary Science of Addictive Junk Food. *The New York Times Magazine*, 20 February.
Patel, Raj. 2010. *Stuffed and Starved: The Hidden Battle for the World Food System.* Brooklyn: Melville House Publishing.
Pollan, Michael. 2006. *The Omnivore's Dilemma: A Natural History of Four Meals.* New York: Penguin Books.

Thomas R. Sadler
Macomb, Illinois
February, 2013

Chapter 1: Introduction

Thomas R. Sadler

In March, 2009, The Illinois Local and Organic Food and Farm Task Force published a report, "Local Food, Farms and Jobs: Growing the Illinois Economy," that argued something very specific. While Illinois consumers spend $48 billion annually on food, almost all of this money leaves the state. Retaining a larger share of Illinois food dollars would create jobs, civic engagement, and a greater connection between food producers and consumers. For these activities to occur throughout the state, the demand for and supply of locally-sourced food must increase.

Across the state, farmers markets and community supported agriculture (CSA) programs continue to provide a greater share of the state's food supply. Currently, not enough locally-sourced food exists to meet growing demand from grocery stores, restaurants, hospitals, schools, colleges, universities and corporate kitchens. To meet this demand, a new food supply chain must emerge. The supply chain must shorten the distance between the farmers' fields and the consumers' plates. A new and expanded supply chain requires changes in the way most people think about the economic, political, social and cultural institutions that govern food networks (Illinois Local and Organic Food and Farm Task Force, 2009).

A change in something as important as our food supply chain will involve elements of public policy and private initiative. In the context of public policy, the legislature may encourage state institutions to procure a greater percentage of their food locally and eliminate regulatory barriers that restrict local food production and marketing. But most of the change must stem from the demand side of the market, by individuals deciding they want to spend a greater percentage of their budgets on locally-sourced food.

Incentives

Economists teach that people respond to incentives (Schiller, 2013). When society properly crafts economic incentives, producers will respond in such a way as to boost specific areas of local food production. The aforementioned growth in farmers markets and community supported agriculture programs in the state of Illinois serves as a case in point. In their chapter in this book on the growth of CSAs, Christopher Merrett and his co-authors, all members of the Illinois Institute of Rural Affairs at Western Illinois University, explain that the increase in interest in local food stems from concerns of environmental degradation, food safety and quality, and community development.

In the context of local food production, consumption, and activism in central Illinois, this book addresses these and many other examples. An ongoing theme in the book is that, in our region, locally-sourced food currently serves as a viable alternative to processed food. But, for locally-sourced food to continue to increase as a percentage of the total food supply, changes must occur. By choosing to consume locally-sourced food, households must demand these changes. Farmers and other entrepreneurs must build local supply chains capable of delivering larger volumes of Illinois food to the marketplace, and not just corn and soybeans. In this transformation, information concerning the availability of local food and the community and health benefits of eating locally must continue to spread. Individuals and families must make a conscientious effort to seek out local food, cook, savor the delicious nature of home-cooked meals, and value the fellowship that comes from eating with others.

In the process of expanding the local food network in our region, many roadblocks exist. County-level health agencies may prohibit the sale of locally-sourced food outside farmers markets. Elementary schools may resist the planting of gardens, because time and resources are scarce. Restaurants may not adopt seasonal produce on their menus because of a lack of convenience in receiving intermittent shipments.

Nevertheless, many stories of progress fill the pages of this book, including backyard gardens that provide surplus vegetables to farmers markets, a community garden that includes members of different socio-economic backgrounds, a seed exchange that strengthens social relationships, CSAs that build community as well as healthy appetites, and a successful fruit farm and creamery that sells output to local and regional markets.

Local and Sustainable

Dr. Joel Gruver, Assistant Professor of Agriculture at Western Illinois University, states in an interview in this book (Chapter 3) that the simple definition of sustainability is "an approach to agriculture that actively attempts to address social, economic and environmental concerns." But not only has the growth in local food networks raised awareness about the health benefits of eating local fare such as grass-fed beef and organic vegetables, clear economic, social and environmental benefits also exist. These latter benefits include job creation in agricultural production, the reduction of chemicals in the soil, greater landscape diversity with perennial and annual planting, a heightened sense of community resulting from all-inclusive local food dinners, a community garden, a common pickup time and location for a CSA, and an emphasis on cooking with fresh and locally-sourced ingredients.

However, determining whether the growth of a local food network is sustainable requires specific metrics that gauge the social, economic and environmental outcomes. Steven C. Hackett (1998), in *Environmental and Natural Resource Economics*, argues that sustainability is a normative standard or a vision of the future, arguing that it "calls for a more sophisticated and inclusive view of economic development and well-being that explicitly takes into account ecological health, natural resource stocks, and aspects of social justice." There is no doubt these aspects of sustainability must form the basis for a local food network. But judging progress with respect to ecological health, natural resource stocks and social justice requires a careful examination over time of specific metrics that measure these outcomes.

Such a local food network "scorecard" must take into account the different kinds of community capitals that build local assets for positive community change. Research in this book on the wisdom and vitality of the local food movement (Chapter 4) demonstrates the challenge of developing such a scorecard. In particular, if goals of the development of a robust local food network include the creation of a vital economy, social inclusion, and movement toward a healthy ecosystem, it is important to establish a link between collective efforts of individuals participating in the local food network with specific community outcomes. To analyze this link, consider two areas of research: community development and ecosystem health.

In terms of research on community development, Flora and Flora (2008), in *Rural Communities: Legacy and Change*, argue that communities most successful in developing sustainable institutions and long-term processes of economic development build seven types of community capital: natural, cultural, human, social, political, financial and built. Consider the meaning of each. Natural capital refers to assets that abide in location, such as amenities, soil and water. Cultural capital means the way people understand the world and how to act within it. Human capital reflects the skills and abilities of individuals to contribute to economy and society. Social capital includes the connections among people and organizations and bonds that create community cohesion. Political capital constitutes the abilities to influence rules, regulations and enforcement. Financial capital becomes available when businesses support a community's capacity for entrepreneurship and investment. Built capital represents the infrastructure that supports the community, including water, telecommunications and roads (Flora and Flora, 2008).

This framework provides a method in which a community may judge progress, but the question of the establishment of specific metrics remains. For example, how should a community evaluate changes in natural, cultural or social capital? Even though a community may judge progress in its built or financial capital by focusing on the quality of the infrastructure or changes in per-capita income, political capital may remain elusive. Do individuals have a voice in community or regional governance? Are the decisions of local government adequately moving the region forward in terms of a local food network? Are the voices of community activists heard? In addressing these questions, which are case-specific and dependent on the economic and institutional environment, research on ecosystem health provides context.

In 2007, the journal *Ecological Economics* published a special issue on "Ecosystem Services and Agriculture." Papers in this issue address such topics as market-based approaches to providing ecosystem services, spatial incentives to coordinate contiguous habitat, ecosystem services and rural poverty, and many other topics. For the purpose of establishing a framework to judge progress in a local food network, three articles from this issue merit particular attention.

First, Dale and Polasky (2007) establish a framework to evaluate the outcomes of agricultural practices and ecosystem services. They argue that the environmental impacts of agricultural practices have consequences on ecosystem services such as water quality, pollination, carbon sequestration, soil retention and biodiversity conservation. In turn, these ecosystem services impact agricultural services. This "circular flow" reveals the informed choices with respect to the most beneficial agricultural practices.

However, challenges of system complexity and metric efficiency exist with respect to the development of a suite of indicators for agricultural practices. What metrics adequately measure changes in a complex agricultural system that are simple enough to model and monitor? We may answer this important question in the context of a local food network, where it is possible to identify specific food producers and markets. An increase in the sales of locally-sourced food would represent a positive outcome. As more local and organic produce flows into a local food network, the quality of ecosystem services increases.

In another paper from the special issue of *Ecological Economics*, Swinton *et al.* (2007) offer additional insight. They determine that agriculture provides and receives ecosystem services that extend beyond the provision of food, fiber and fuel. Services that flow to agriculture include climate and air regulation, pollination, pest regulation and genetic diversity. Services from agriculture include landscape aesthetics, recreation, carbon sequestration and biodiversity conservation. Swinton *et al.* argue that, given this framework of agricultural services among managed ecosystems, properly crafted agricultural practices have the potential to diversify the suite of ecosystem services. That potential arises from the objectives of human management practices. In particular, sustainable agricultural practices may diminish the reliance on external agrochemical inputs. Farmers may alter surrounding habitats in a landscape to enhance the productivity of an agricultural system.

In a third paper from the special issue of *Ecological Economics*, Zhang *et al.* (2007) argue that managing agricultural landscapes to enhance ecosystem services requires collaborative and multi-disciplinary research. In particular, supporting ecosystem services (ES) such as soil structure and fertility and regulating services such as natural control of plant pests contribute to the strength of agricultural ecosystems. Ecosystem dis-services (EDS), such as habitat loss and nutrient runoff, reduce the productivity of agricultural ecosystems.

To understand long-term impacts, it is necessary to determine the scales in which ES and EDS are normally provided and the activities that typically supply them. As one example, the ecosystem service of soil retention is managed by the farmer with cover crops; however, the landscape surrounding the farm, consisting of vegetation or a floodplain, also contributes to soil retention. Other examples of the immediate and surrounding ecosystem services include genetic diversity, pest control, and water provision and purification (Zhang *et al.*, 2007).

These three papers provide context for an important question: how should a community judge the progress of a local food network? That is, if the establishment of a local food network impacts the levels of community capital—including natural, cultural, human and social—how may we determine if the change is positive? To be sure, we need "simple metrics that identify opportunities for improvement," as Dr. Joel B. Gruver states in Chapter 3.

The development of such a framework would collect the important outcomes of a local food movement into one scorecard. These aforementioned papers from *Ecological Economics* discuss the environmental benefits of properly-crafted agricultural practices. For the local food network to thrive, consumers must demand environmental quality in addition to healthy food. These consumers must choose to build, not destroy, natural capital. In this context, the symbiotic interaction between the provision of local food and sustainable use of the land provides a conceptual way to justify activism in the local food movement.

It is important to broaden the context of the papers in the special issue of *Ecological Economics* to consider all capitals impacted by a thriving local food network. As Chapter 4 discusses, the outcomes of a growing food network offer specific areas in which to judge progress: econetwork services (food, soil retention, nutrient recycling), cultural services (agricultural lifestyles, aesthetic landscapes, spiritual well-being), externality elimination (higher air, water and soil quality), and community development (democracy, institutions, policy). Using these and other metrics, a community or region may establish whether a local food network is contributing to the goals of growth in the demand and supply sides of the local food network, the creation of economic opportunity in the form of jobs, and the improvement of environmental and social conditions.

Impediments

Two important reasons reveal why local food networks continue to provide a small percentage of the food supply. First, over the course of the last six decades, industrial agriculture, farm policy and a national food chain have increased the efficiency of production and distribution of meat and processed food to the extent that our society faces an "omnivore's dilemma" of what to eat, not how to obtain enough calories (Pollan, 2006). The supermarket food available to generations of families since the 1950s moved society away from local food networks. In the context of recent history, the convenience and supply of food produced by the industrial food system are important accomplishments, despite the negative externalities associated with this system: pollution, environmental degradation and negative impacts on human health.

Second, local food networks, starting from the activist efforts of small groups of citizens, often reach their point of saturation among consumers in a community and then struggle to expand beyond the original group. For the majority of people that shop in traditional grocery stores and consume industrial food, why make the extra efforts to purchase locally-sourced alternatives and develop the necessary cooking skills? Because of the ubiquitous nature of processed and fast food, these efforts are not necessary.

But other challenges remain. In central Illinois, expanding local food networks will remain a fundamental challenge for the foreseeable future. As A. Bryan Endres, Lisa Schlessinger and Rachel Armstrong argue in Chapter 9 on local food networks and the legal system, communities choose the rules and regulations that encourage or prohibit what the communities collectively value. But related federal, state and county laws concerning the food system may complicate this process of oversight. In fact, conflicting rules at different levels of government may restrict the expansion of local food systems. One attempt to address these issues, as Endres, Schlessinger and Armstrong explain, is Cottage Food laws, which exempt certain small food producers from inspection and licensing requirements in the Food Code. These laws vary by state, but apply to producers with household operations and small-scale production as defined by gross sales.

As another example, growth in the local food movement is synonymous with the strengthening of the local food chain. But, in the case of local produce, the food chain starts with the planting of seeds. Soil science and organic growing techniques inform this important process, but the farmer or gardener must first obtain seeds. As Molly E. Hicks elegantly argues in Chapter 8 on seed exchanges, the procurement of seeds within sustainable agri-food networks constitutes a fundamental aspect of a local food chain. But most small-scale growers in the heartland and now worldwide obtain their seeds from outside seed companies, where the seeds are grown under different growing conditions.

The farmer or gardener who purchases seeds from outside sources may not know how the seeds were grown, when they were harvested or the conditions of processing and transportation. These issues are the same ones that disconnect consumers from producers when purchasing food in traditional grocery stores. As a result, as Hicks explains, seed exchanges that establish networks for individuals to give, sell or trade seeds strengthen economic and social relationships while providing a setting for the exchange of farming knowledge.

Another challenge in expanding the local food network besides the legal system and the establishment of a sustainable local food chain is attitudes toward cooking and human health. Individuals who want to limit the consumption of industrialized foods must not only seek out locally-sourced alternatives, but they must allocate time and energy in the kitchen to cooking. The processes of both purchasing locally-sourced food and building a knowledge base to cook with fresh and local ingredients require time and effort. For generations, this knowledge was passed on from mother to daughter (and sometimes to son). But the growth in the industrial food economy and the higher labor force participation rate of women since the 1960s largely eliminated this cooking tradition.

For many households, work responsibilities and a commitment to kitchen time may conflict. However, as Holly A. Stovall and Jamie Lane explain in their chapters on cooking and nutrition, health, environmental care and the freshness of food all serve as arguments for the consumption of local and chemical free foods that are grown by environmentally conscious farmers. Most fruits and vegetables that come from the system of industrial agriculture are exposed to fungicides, herbicides and pesticides and sometimes fecal contamination, if livestock are nearby. These contaminants may cause chronic health problems. Unless individuals and households value a healthier way to eat, chemicals, additives and pollutants will accompany the consumption of many industrial food items.

A final example of an impediment to the growth of a local food network is a lack of community activism. If people want locally-sourced food, they must make the effort to grow it themselves or buy it from local food producers. Fortunately, in our region, local food options are becoming more plentiful and diverse.

Heather McIlvaine-Newsad and Rob Porter, authors of Chapter 12 on Macomb's Community Garden, cite several reasons why they, along with Joel B. Gruver, started and continue to participate in this project. First, they believe in social justice. Access to healthy produce should be available to all. Second, sharing growing tips and conversation during the processes of planting, weeding, watering and harvesting for people of different socio-economic backgrounds strengthens a community's social capital. Finally, activism in the context of a community garden expands the local food network.

Seizing the Opportunity

Farming and thriving local food networks may serve as economic engines for entire states. The 12.8 million consumers in Illinois that purchase processed food from restaurants and supermarkets export tens of billions of dollars to California, Idaho, Mexico, China and other areas across the country and around the world Illinois Local and Organic Food and Farm Task Force, 2009). But thriving farmers markets in Springfield, Champaign-Urbana, Bloomington-Normal, Galesburg, Macomb and many other communities, along with the growth in famous CSAs such as Barefoot Gardens in Macomb and Henry's CSA in Bloomington-Normal, demonstrate that many people demand locally-sourced food options. Statewide, almost 200 more outdoor farmers markets appeared over the course of the last decade (Illinois Local and Organic Food and Farm Task Force, 2009).

In addition, as awareness increases of the environmental costs of industrial food, the benefits of eating healthier alternatives and the merits of bridging the gap between farmers and consumers, more locally-sourced food options will emerge. With industrial food, the cost of shipping may account for more than 10% of the price that consumers pay (Illinois Local and Organic Food and Farm Task Force, 2009). A thriving local food network that establishes an efficient system of distribution, however, may reduce or eliminate that energy and distribution surcharge. This savings partially or fully offsets the premium many consumers are willing to pay for organic or naturally grown alternatives. Using its competitive advantage of fertile soil and agricultural knowledge, rural areas in central Illinois and throughout the state would benefit economically by feeding their urban neighbors (Illinois Local and Organic Food and Farm Task Force, 2009).

Ultimately, the process of creating and sustaining all links in a local food supply chain—planting, growing, harvesting, distributing and consuming—translate into economic opportunity in the form of jobs. The community multiplier demonstrates that as consumers spend more money locally, the overall impact is a multiple of the original transaction. Therefore, seizing the opportunity to build on the momentum that already exists in the local food movement would create both short-term opportunities in the context of more delicious and local food options and the long-term benefit of building sustainable local food networks.

Networks as a Paradigm for Development

Addressing the local food movement in central Illinois in the context of networks provides utility in understanding diverse forms of economic development. Any analysis such as this must consider the ways new network forms interact with pre-existing socio-economic institutions. Successful regions emerge in the areas that possess a diverse set of economic and social forms (Murdoch, 2000).

To understand why the concept of networks provides a model to address the local food movement, consider why knowledge in a particular area—such as local food production or consumption—may be viewed as a product of a network of heterogeneous materials. The system or marketplace that emerges over time from the interaction of a number of food sellers, each with personal knowledge of planting, growing, harvesting, distribution and selling, and a number of food buyers, is the end product of individual choices and behaviors, hard work, and the experience of heterogeneous agents and actions.

In an important article on actor-network theory, Law (1992) describes this process as the accumulation of efforts when the "social, the technical, the conceptual, and the textual are fitted together" into a "patterned network which overcomes their resistance." In the local food movement, a patterned network of heterogeneous materials shapes the social and economic interactions that lead to growth (or decline) in the movement. Therefore, the structure of a network is not static, but is the site of choices and change, of building and tearing down, and exists as a "relational effect that recursively generates and reproduces itself" (Law, 1992).

The important implication of this idea of a patterned network is that new centers of economic organization may emerge. The pre-existing institutions of influence (corporations in a capitalist free-market economy and farmers in a local food economy) are contestable from the perspective of individuals who want a more sustainable system of production and consumption. The challenge is how community actors juxtapose, mobilize and bring together the characteristics of a local food market, prevent these characteristics from following their own ideas and beliefs, and turn a gathering of individual parts into both a network of influence and the institutions of a viable marketplace.

As a way to apply this theory of networks to the local food movement, Murdoch (2000) considers major areas of study of the modern food network: the production process, the organization of food production, the use and management of labor, the role of research and extension activities, and marketing and distribution. Murdoch argues that these areas of study highlight the nature of the "commodity chain" that constitutes the modern food industry. But these aspects of the commodity chain also reveal how the components interlock and influence the individuals who participate in the different economic processes. Large-scale industrial companies dominate the modern commodity chain. This reality has given rise to factory feedlots, global systems of distribution, and a focus on economies of scale. As a result, the food industry has substituted natural processes with their industrial counterparts, disadvantaged traditional rural residents, and degraded the environment (Murdoch, 2000).

Longer food chains increase their socio-economic complexity. But local food networks are constructed and maintained by the natural and social actors that create them. These shorter food chains eliminate many areas of socio-economic complexity, including large fossil fuel requirements, long distribution networks, the chemical treatment of soil, the food-like additives that serve as resource inputs in the processing of industrial food, and many other factors.

The socio-economic constituents who hold the local food network together are part of a wider movement to strengthen the place of the farmer. Different food networks demonstrate alternative methods of valuing outcomes of food production and consumption. In local food networks, issues of food quality, environmental and human health, and sustainable production techniques form the basis of the paradigm. As a result, the actions of all actors involved in the local food marketplace—on both the demand and supply sides—state preferences for ecological integrity, economic vitality and social opportunity.

An important part of this analysis is consideration of production-consumption relations. Traditional analyses, or Systems of Provision (SOPs), demonstrate the commonly understood vertical integration of a food system (Lockie and Kitto, 2000). The dimensions of production, distribution, retailing and consumption are all important in the context of the material culture surrounding them. But a theory of vertical integration with respect to local food systems highlights consumption at the end of the chain. In this manner, consumption practices may be denied any status of delineation, as multiple commodities are incorporated into consumption practices. As a result, the stages of production, distribution and retailing serve as determinants of consumption behavior, while these variables do not actually exhibit causal relationships with consumption (Lockie and Kitto, 2000).

The reason this framework is lacking when describing local food networks is that consumption constitutes a variable as important as any on the supply side of the market. The SOPs analysis does not address the complexity of consumption practices.

Another framework is better suited to describe the concept of local food networks: Actor-Network Theory (ANT). This framework acknowledges the interconnectedness of the social systems and actors in a setting of economic activities. Analyses of networks in this context make no *a priori* assumptions concerning what actors or agencies should be attributed. With this method, results occur both theoretically and empirically. An emphasis on local actors and agencies in shaping food processes means an analysis of local food networks requires observation of these important constituents (Lockie and Kitto, 2000).

Laura T. Raynolds (2002) develops this idea further. She argues that commodity frameworks are commonly used to describe the problems of industrial food systems and potential benefits of alternative systems. In a revised commodity approach to food networks, the contributions of actors and activities strengthen the traditional supply chain framework. This is important in the description of local food networks as many are "buyer-driven" (Raynolds, 2002).

Consumer knowledge and behavior may mobilize local production and consumption. In this context, the knowledge of household cooks may reflect and demonstrate actions to shun processed food. An effort to link consumption and production reflects the complexity of social relations connecting many farmers, gardeners and consumers. According to Raynolds (2002), this network analysis focuses on the economic, social and cultural relations that form "dense webs of informational and material interdependence" and addresses two issues:

- How local agents extend their reach via other humans and technologies, and

- How networks woven by practices and capacities persist across time and space.

In an article that brings this material together and offers a useful way to think of the characteristics of a local food network, (Morris, 2004) explains that ANT uncovers ways in which particular models gain acceptance in custom and practice. With respect to a local food network in a particular region, what is important is how particular processes, ideas and categories come into existence and how they are maintained. The actors and agencies form a local food network by virtue of the fact that they work with others and share similar goals and values. The heterogeneous materials that Law (1992) writes about is relevant in the sense that producers, consumers, agencies, markets and technologies are brought together with the capacity to act. This "relational materialism" enables networks to stabilize and extend across time and space. Therefore, the combination of the social and the material sustain the economic interactions of a vibrant and sustainable local food network (Morris, 2004).

Framework for a New Food Network

A framework for a new food network must include the idea that consumption decisions are as important as sustainable production decisions. A symbiotic relationship between producer and consumer insures a market-clearing quantity of agricultural output. In their chapter on the origins, successes and contemporary challenges of a fruits farm and creamery in central Illinois that sells to local and regional markets, Wesley Jarrell and Leslie Cooperband make the case for such a network. According to these former academics and current business owners, a new and sustainable food network should include the following characteristics: a holistic approach, a focus on health and food quality, economic and landscape diversification, human-and-animal connections, and energy and water biodiversity.

In reality, these social, ecological and economic circumstances continue to change, according to market and environmental impacts. Nevertheless, a holistic approach to the land and the marketplace demonstrates a long-term commitment to quality and variety. In a new and sustainable food network, producers must adapt to dynamic circumstances and adjust their winter and spring planning decisions, according to previous and anticipated market sales. Consumers must stand ready to take action to influence current and future market outcomes.

The network aspect of this framework reveals agricultural methods that enhance, first, the health of the soil, and, second, the health of human communities. This aspect serves as a helpful solution that, in the words of Wendell Berry (1981), "improves the balances, symmetries, or harmonies within a pattern." Thus, a new and sustainable local food network that solves for pattern in Berry's framework also solves for more than one problem and does not create new problems. What are some of the outcomes of such a desirable system? That is, what problems or challenges are addressed by a well-functioning local food network?

In her engaging book, *The Seasons on Henry's Farm*, Terra Brockman (2009) answers these questions in the context of a year of food and life on a sustainable farm. As the sister of Henry Brockman—who operates both a community-supported agricultural program in the Bloomington-Normal area in central Illinois and a stand at the Evanston, Illinois farmers market—and the founder of The Land Institute, Brockman writes with knowledge and experience about sustainable food production.

By tracking the activity that occurs during the twelve phases of the moon on Henry's Farm and discussing the market exchanges that result from Henry's labor, Brockman (2009) demonstrates the "hunger for connection" that many people have to the "source of all sustenance, the earth, and the people who work it." She demonstrates how all things human and agricultural interrelate, and how all humans, when we eat, become part of the cycle of life and death via the animals and plants that provide our daily sustenance. But *The Seasons on Henry's Farm* also addresses Wendell Berry's question of how appropriate farming techniques and systems solve multiple problems simultaneously. Consider a few examples from Brockman (2009):

- Small farms may serve as economic and environmental prototypes.
- Local outlets for farm produce connect consumers to the growing process.
- Small and sustainable farms and local food markets demonstrate the interrelatedness between ecological, social and economic processes.
- In a local food network, a symbiotic relationship exists between producers and consumers.
- A sustainable farm may exhibit a diversity of landscape.
- A small and sustainable farm demonstrates Aldo Leopold's idea of a "land ethic."
- The challenges of long-term sustainability are ongoing.
- Locally-sourced food production and consumption attempt to address the challenge of food security.
- Seasonal eating constitutes a crucial element in a local food network.

The concern for accelerating carbon emissions and commercially raised produce shipped thousands of miles makes Henry's Farm and other forms of local food production economic and environmental prototypes for local resilience. In addition, the hunger for connection to the growing process that many consumers demand is addressed through the local and transparent methods that accompany CSAs and farmers markets. A demonstration of ecological, economic and social interrelatedness, necessary for the development of communities and sustainable food systems, reveals how food systems may enhance the economic viability of a region, not destroy it. Transparent and accessible agricultural production techniques, interaction between market participants and feedback from customers creates a simple and efficient food network and a symbiotic relationship between producers and consumers. Market interaction leads to diversity in planting. In a sustainable farm, low-gross-revenue crops provide diversity to the agricultural process, complementing output with greater market share. A diverse array of crops hedges against environmental and market problems that result from recessions and droughts (Brockman, 2009).

Perhaps the most important aspect of a new and sustainable network of local food options is the realization of the "land ethic" of Aldo Leopold (1949) as a cooperative and extended kinship that "enlarges the boundaries of the community to include soils, waters, plants, and animals, or collectively: the land." An appreciation of the land and its role in the vitality of a community and region entails a system that focuses on the web of life between humans, plants, animals, weather and soil, long-term sustainability, and food security. In this framework of a vital local food network, some farming practices are better suited than others. According to Brockman (2009), the best farming practices adhere to Leopold's land ethic.

An important outcome of a symbiotic relationship between producers and consumers, a sustainable land ethic, and a focus on locally-sourced food is the reality of seasonal eating. While kitchen mainstays such as coffee, sugar and salt lead to global supply chains, many food items consumed in the heartland are currently grown in this region. Tomatoes, peppers, edamame, squash, beans, cucumbers, spinach, potatoes, eggplant, carrots, salad greens, herbs and, of course, sweet corn form the basis of many household diets during the months of harvesting; however, a growing number of green houses and hoop houses in the area provide these and other forms of produce for additional months throughout the year.

Nevertheless, the availability of oranges in January in Illinois serves as one example of the miracle of the global food system. But should we choose to consume all food items all the time without regard to environmental and logistical constraints? If we choose to become more selective and local, then we won't have all food items available at all moments in time. Familiarity with the growing process reveals the fact that the pendulum swings between abundance and frugality, depending on the season, the weather, growing and harvesting techniques, and many other factors. Terra Brockman (2009) once again explains this process elegantly, referring to the month of June when peas are harvested and eaten:

> Such pea picking and freezing marathons epitomize the link between food supply and season. If fall and winter teach us to be conservative, economical, and restrained, then late spring and summer teach us to be generous, openhanded, and free. The fact that every crop is at its peak for only a short time promotes a spirit of making the best of it while it lasts, and conserving part of it for future use, as good a metaphor for how to live one's life as I have ever stumbled across.

That we choose to self-limit the variety of food we consume according to seasonal constraints is fundamental to the concept of sustainable food networks. In the context of the aforementioned outcomes of a new and sustainable food network that satisfies Leopold's land ethic, this may be the most important. The industrial food system's efficiency in distributing output to markets around the world is nothing short of monumental. By choosing to consume a greater level of locally-sourced food, however, we support the producers that we know and help develop the symbiotic relationship between farmers and those who adopt nutritious diets.

A Model of Success

While the books and articles of Wendell Berry have long inspired local food advocates, many examples exist of cities and regions that have embraced locally-sourced food and developed the networks that provide economic opportunity. Ben Hewitt (2009) writes of one example: Hardwick, Vermont. Arguing that we need to rethink our entire food chain for reasons of health and economic security, Hewitt makes the case that many residents of the Hardwick region are attempting to create a vibrant local food network. Because this region in Vermont has a history of small dairy farms, the large-scale production operations that include hundreds of acres—common in the Midwest—never took hold in Vermont. As a result, many farmers and other entrepreneurs have established businesses that serve local and regional markets, including a seed company, cheese business, restaurant and bar, center for composting and many others.

Because a centralized food system is vulnerable to food-borne illnesses and energy-supply disruptions, a local food network guards against these harmful outcomes. Hewitt (2009) explains that, because of the circular nature of a local food network, the goals of the network include the creation of economic vitality, food production from solar energy, and local nourishment, whereby the:

> measure of a local food system is only partly the system itself. You have to judge local food systems in the context of where they are. By this reckoning, a healthy food system will always be the one that's moving toward the lower end of whatever scale is adequate to keep the locals nourished, spiraling toward the shortest, least-complex supply chain necessary to ensure that food reaches table…the appropriateness of any agricultural endeavor can't be measured merely against the extremely poor health of our industrial food system, it also must be measured against whatever alternatives already exist.

In central Illinois as in the Hardwick, Vermont region, locally-sourced food alternatives already exist and continue to grow in importance. While the small-scale nature of farming in Vermont lends itself well to a local food network, the knowledge and tradition of agriculture contribute to the growth of the local food network in central Illinois. We may not have Ben Hewitt to popularize the local food movement in our region, but the authors in this book argue that the local food movement here is vibrant and growing.

Global Concerns

While the context of this book is central Illinois, many reasons of global significance exist as to why the development of local food networks is important. On a global scale, population growth is slowing, but aggregate population continues to rise. Land is fixed. Therefore, in order to increase profits, the industrial marketplace requires higher levels of productivity, more efficient global supply chains, and greater sales of food. This model of globalization is vulnerable to environmental, economic and energy shocks. The world continues to experience examples of all three.

In the summer of 2012, in North America, scorching heat and a severe drought increased food prices, leading to worries about a rise in global food costs (Lowrey and Nixon 2012). The drought reduced much of the 2012 corn crop, increasing the price of processed foods, ethanol and animal feed. The level of global integration of commodity markets—including food—makes it possible for regional disruptions such as this drought to impact people around the world.

With globalization, the expansion of the biofuels industry—dependent on the availability of land and vulnerable to environmental shocks such as drought—contributes to spikes in food prices (Rosenthal 2013). In particular, the increasing use of Midwestern land in the United States to grow corn for ethanol production reduces the supply of corn for processed food. This market reality puts upward pressure on the prices of processed food. In addition, recent laws in the United States and Europe mandate the increasing use of biofuel in cars. These two economic realities impact people throughout the world.

In her elegantly written book, *Diet for a Hot Planet*, Anna Lappe (2010) argues that industrial agriculture serves as an important contributor to climate change. The recent weather-related global problems that we've observed—droughts leading to crop losses, soils ruined by salt accumulation, flooding that leaves soil waterlogged, longer growing seasons bringing new and more pests, and erratic weather shifting harvest seasons—demonstrates that the global system of food production and distribution is intimately intertwined with global warming. In fact, according to Lappe (2010), the "food sector is a critical part of the problem." The reason is the reliance on international supply chains that demand fossil fuels, systems of monoculture that lack crop and animal diversity, and a belief among the public that we consume all types of food at all times. In contrast, implementing climate-friendly solutions with respect to the food system creates "beneficial ripples," including the preservation of biodiversity, the enhancement of food security and human health, a lower reliance on fossil fuels, and stronger communities (Lappe, 2010). Local food networks and activism in the heartland and throughout the world may serve as important components in addressing this global environmental problem.

Conclusion

The contributors to this book hope that the original research and writing in the following chapters will help shape how readers think about the potential of local food networks. While the benefits of strong local food networks are clear—establishing a link between producers and consumers, strengthening communities, providing healthier food options for cooking, improving environmental quality, reducing the reliance on fossil fuels, providing farmers and gardeners who want to grow a diverse set of crops a more robust marketplace, and many others—it may be that many people will ultimately decide to produce and consume locally-sourced food for moral reasons. In Chapter 13 on the moral foundations of the local food movement, Brian K. Powell argues that the moral heart of the local food movement is about knowing the story of where our food comes from. For example, the relationships that consumers cultivate by picking up their weekly basket at John Curtis' Barefoot Gardens CSA in Macomb, or purchasing cheese from Wesley Jarrell and Leslie Cooperband's Prairie Fruits Farm and Creamery in Champaign, or buying from one of a number of farmers markets in central Illinois enhances the consumer knowledge base. As a result, consumers know where their food comes from, how it is grown, if the animals they are eating are treated humanely, and whether their participation in a food network contributes to community vitality. The consumers know that through their purchases they are providing John, Wesley and Leslie with the opportunity to earn income by farming.

These local food producers are not nameless cogs in an industrial process. They are community members who are making thoughtful and difficult decisions with respect to agriculture, economics and the environment. Despite the obstacles they face, they have contributed much to the local food economy. Because their contributions and the contributions of others have improved our region in many ways, we hope our readers will appreciate the significance of the local food movement in the heartland.

References

Berry, Wendell. 1981. Solving for Pattern, in *The Gift of Good Land: Further Essays Cultural & Agricultural*, North Point Press.

Brockman, Terra. 2009. *The Seasons on Henry's Farm*, Chicago: Surrey Books.

Dale, Virginia H. and Stephen Polasky. 2007. Measures of the effects of agricultural practices on ecosystem services, *Ecological Economics* 64:286 – 296.

Flora, Cornelia B. and Jan L. Flora. 2008. *Rural Communities: Legacy and Change*, 3rd edition, Westview Press.

Hackett, Steven C. 1998. *Environmental and Natural Resource Economics*. Armonk: M.E. Sharpe.

Hewitt, Ben. 2009. *The Town that Food Saved*, New York: Rodale Books.

Illinois Local and Organic Food and Farm Task Force. 2009. Local Food, Farms and Jobs: Growing the Illinois Economy, A Report to the Illinois General Assembly, March.

Lappe, Anna. 2010. *Diet for a Hot Planet*. New York: Bloomsbury.

Law, John. 1992. Notes on the Theory of the Actor-Network: Ordering, Strategy, and Heterogeneity, *Systems Practice* 5(4): 379-393.

Leopold, Aldo. 1949. *A Sand County Almanac: And Sketches Here and There*, Oxford: Oxford University Press.

Lockie, Stewart and Simon Kitto. 2000. Beyond the Farm Gate: Production-Consumption Networks and Agri-Food Research, *European Society for Rural Sociology* 40(1): 3-19.

Lowrey, Annie and Nixon, Ron. 2012. Severe Drought Seen as Driving Cost of Food Up. *The New York Times*, 25 July.

Murdoch, Jonathan. 2000. Networks – a new paradigm of rural development?, *Journal of Rural Studies* 16: 407-419.

Pollan, Michael. 2006. *The Omnivore's Dilemma: A Natural History of Four Meals*. New York: Penguin Books.

Raynolds, Laura T. 2002. Consumer/Producer Links in Fair Trade Coffee Networks, *European Society for Rural Sociology* 42(4): 404-424.

Rosenthal, Elisabeth. 2013. As Biofuel Demand Grows, So Do Guatemala's Hunger Pangs. *The New York Times*, 5 January.

Schiller, Bradley R. 2013. *The Macro Economy Today*, 13th edition. New York: McGraw-Hill.

Stallman, Heidi R. 2011. Econetwork services in agriculture: Determining suitability for provision by collective management, *Ecological Economics* 71:131 – 139.

Swinton, S.M., F. Lupi, P.G. Robertson, and S.K. Hamilton. 2007. Econetwork services and agriculture: cultivating agricultural econetworks for diverse benefits, *Ecological Economics* 64:245 – 252.

Zhang, W., T.H. Ricketts, C. Kremen, K. Carney, and S.M. Swinton. 2007. Econetwork services and dis-services to agriculture, *Ecological Economics*, 64:253 – 260.

Chapter 2: Sustainable Local Food: Turning the Soil, Turning the Tide

Bill Knox

I'm not a very good gardener. Over the years, I've tried many times but only in the last few years—and with a move to the rich soil of west central Illinois—have I been able to get more out of the ground than I've put into it. But regardless of what happens with the radishes—my favorite—tomatoes, peppers, melons, zukes, and beets, the bigger mystery—and the one that bears on sustainable local food production, is why I care about growing some of my own food and how I want to help others grow their own.

It is true that I have an interest in where food comes from, how to feed people both nearby and around the world who do not have access to adequate nutritious food, and why the distribution of "food power" is controlled by the handful of brands found on our kitchen shelves, but only recently have I realized both the cause of my underlying interest and the need to encourage others to increase their control of food production—to bridge the gap between less-sustainable large-scale food production and more sustainable small scale gardening.

Apart from my long-term interest in sustainability issues, including population growth, pre- and recycling, energy conservation and alternative energy, people who have known me for a long time might say the urge to grow things and to find green things to eat is in my genes: Both of my parents spent at least part of their childhoods either growing up on Midwestern farms or living with parents who had vegetable gardens. These gardens were not hobbies, but were an essential part of feeding large families in the 1920s through the 1940s. Later, one grandmother, perhaps out of habit, maintained a modest backyard garden complete with the rhubarb I munched on as a kid until she passed away in the mid-1950s. My parents, however, settling before World War II in a northern Chicago suburb, never kept a vegetable garden, perhaps because it was "old-fashioned" and because the demands on my dad as a small business owner and my mom as a housewife and bookkeeper gave them little free time.

Nevertheless, my first attempt at gardening came when I was eight or so when I begged my dad to plant some radishes. My first garden was a wooden slat Florida grapefruit box he filled with soil and short rows of radish seed which, eventually, produced some red, bitter, marble-sized roots that I was nevertheless proud of—having watered them for some weeks—but I did not garden again until I was in my twenties.

In the meantime, I was not without four very strong examples of do-it-yourself neighborhood gardeners that always made me wonder at their seemingly effortless edible results. Until I was six in a neighborhood largely built in the 1920s, I lived across the street from Mrs. R.'s family who had built what seemed to be a fantastically long, 1950s brick ranch on a corner lot with almost the entire backyard cultivated in vegetables, flowers, and bordering shrubs. Complementing and supporting this growth were what seemed like miles of brass-nozzled green garden hose, tremendous hoops and trellises supporting tomatoes and beans, and—wonder of wonders—a rototiller bigger than my pedal-powered fire engine. This tremendously lush, well-tended, and mechanized vegetative paradise I watched through Mrs. R's chain-link fence across the street on the north side and a row of arbor vitae and alley on the west. In the full east and south sun, these untouchable veggies grew—it seemed to my five-year-old's mind—to lush proportions—a veritable forbidden post-war Garden of Eden that rivaled the displays in the new supermarkets I shopped with my mom. Years later, dad told me how during those years I would treat him to my make-believe day spent on the farm—probably inspired by equal parts my parents' childhood stories and Mrs. R.'s constant attentions across the street.

Much more accessible were two neighborhood gardens at our next house, where we moved just a few blocks away when I was six. The east side of our yard bordered Mrs. S.'s strip garden extending from behind her garage fifty feet between one neighbor's white picket fence and another's rusty wire one. Each season she would create and tend her garden with long gray, wooden handle spade, hoe, and cultivator. She was the first person I ever saw trowel coffee grounds, fish heads, and assorted kitchen middens right between the rows of bib lettuce, carrots, and tomatoes. Occasionally, we would find a neat pile of fresh, bright green lettuce on our side of our old wire fence that mom would make into wilted lettuce with hot vinegar, sugar, and bacon drippings. Perhaps this nearby example encouraged me to convince Dad that I, too, could grow radishes in box.

Across the street, however, was a place of fresh fruits and vegetables more appealing to a boy of 9 to 11. Behind their bungalow, Mr. and Mrs. F. had a small yard, but one that was divided between the garden at rear near the burn barrel and lilacs and the fruit trees that took up most of the yard. Of course, the chief feature of the pear, black cherry, and apple trees was their low-hanging limbs for climbing. Sometimes, late in the season, the F.s' two grandsons—one a year younger and one a year older than I—would enjoy the ripe fruit without the benefit of washing either it or our hands while sitting on limbs. Mrs. F. tended her veggies while Mr. F. would surprise us during our summer street softball games with the fruits of his labor—homemade root beer he brewed and bottled in his basement near his pot-belly stove and print of the red-faced Jolly Philosopher.

As I look back on those times, I find myself nostalgic and impressed with the self-reliance, nurturing, and generosity of these grown-ups. Although I don't recall Mrs. R. sharing with neighbors, she nevertheless beautified the neighborhood with her meticulous cultivation. Mrs. S. shared her lettuce and company while she worked, intent on pulling weeds or breaking up old subdivision earth. Mrs. F., letting us monkey around in her trees or Mr. F. stamping his foot while playing his cigar box fiddle, invited me into traditions and folkways that were more engaging than our latest sports and war games. As we got just a bit older, we left behind those days of climbing trees and street softball of seventh-inning root beer, but the influence of the fourth gardener was not so easily left behind.

On the edge of town, lived my widowed aunt, my mother's sister, who, even after she sold most of the acreage of the family farm, still gardened, kept chickens, cut firewood, and otherwise did her best to support my cousins living on the margins of life after two lost husbands. I recall hearing of my cousin cutting firewood before she went to school in the morning and my aunt tending her bees and canning, canning, canning from her large, naturally fertilized plot. Even until recent years before she became ill, she would still grow much of her own food and share with family as well as strangers.

Perhaps it is this spirit that stayed with me all these years. When I moved to central Illinois in my twenties to attend graduate school, I was invited to live with my best friend on a dairy farm in a Pullman railroad car set up on a red-brick foundation for a long-past generation of milkers to live in the days before electrification. I don't recall who had the idea, but soon his parents, avid landscape gardeners, brought their talents (and rural backgrounds) to bear on the huge plot behind the LP tank. Our farmer landlord was happy to plow and manure our plot, and my friend's parents brought a small rototiller, seed, plants, and all the assorted tools to us poor students and went to work with us for two seasons of remarkable cooperation and production. I welcomed this opportunity to save grocery money, eat well, and to put into practice the examples of my childhood and teen years. Without any previous training, I learned how to prepare beds, plant, cultivate, and harvest—shared labor coming very naturally with my best friend and his parents. Mrs. K. blanched and froze green beans and Brussels sprouts and canned tomato sauce and. . .more beans! I suppose in some sense we were trying to "get back to the land" as so many twentysomethings did in the early 1970s, but we also had advice from both my friend's parents and our landlady who kept a considerable garden herself just a quarter mile down the road. More important, we grew a considerable amount of food that we ate all winter and shared with our friends.

In the following years, dominated by teaching, more graduate school, and family in Illinois and Michigan, I grew tubs and small plots of tomatoes and other veggies, alone and with friends. But it wasn't until the 1990s I first became involved in a community garden in Michigan's Upper Peninsula. With the encouragement of my preteen daughter and a university colleague who was one of the organizers, we rented a small plot, and made the best of the short growing season. An army of service organization volunteers rototilled the plots near the local cemetery and shareholders put up a deer fence and ran a water line, and later many of us enjoyed a variety of produce and fellowship during long northern summer days.

More recently, with my return to central Illinois almost seven years ago, I have become both more active and reflective about my gardening and the role of local food production relative to the larger issues of resource conservation, energy consumption, and global warming. I have met with gardening success, as I mentioned above, and my family also bought its first share in a local garden. We ate many greens we would not otherwise, including kale, collard greens, and kohlrabi, all of which we enjoyed. More important, the experience started my reading and local inquiries about the local food movement. For the first time the arguments against long-distance transport of domestic fruits and vegetables from California and Florida, international transportation of tropical foods—even "common" foods such as bananas, and formerly "exotic" spices and beverages such as pepper, tea, and coffee, made sense to me. Barbara Kingsolver's *Animal, Vegetable, Miracle*[1] throughout was especially informative, encouraging, and entertaining. How is it that we can be faced with growing planetary scarcity—either in natural or market terms--of cheap petroleum energy while insisting on an eating lifestyle based on foods coming hundreds if not thousands of miles? I began to consider ways to "bridge" the transition from a food economy based on immigrant labor, petrochemical fertilizers, herbicides, and pesticides, and water, land, and "jet fresh" shipping of everyday diet to one over which individuals and communities could not only exert more control but have enriched their lives as well.

A strong thread in my thinking has been the increasing rejection of industrial—mechanized and chemical—approaches for growing, favoring simpler tools and healthier behavior. I have used a larger carbon-footprint approach with a rototiller, lawn tractor and trailer for carrying tools, soil, and compost, and had local soil trucked in for my raised-bed gardens. I have experienced the strong cultural urge since I was a TV kid to mechanize even the most organic of processes. I still use a gas-powered mulching mower on my yard to eliminate the use of municipal yard waste bags and to build soil. But with the garden plot at my new house, it is strictly an earth fork-and-wheel-barrow affair. Although I am approaching retirement, I like more and more the resistance and pace of hand tools for soil work. For stakes and cages, I now build my own from re-bar that will last as near as forever as I can make them. On the chemical side, mulch, compost, and marigolds work to fertilize and protect much of my garden from beetles and rabbits. As I have left aside the labor-saving contemporary "commonsense" approach to gardening with its array of convenience chemicals and gadgets, I believe that I am doing nothing new, but simply following examples I absorbed from the gardeners of my childhood who grew up themselves with no small-scale power tools and few garden chemicals.

It seemed to me that anyone in good health and average intelligence and having access to a bit of private or public land could grow as much or as little food as they would like. However, those also wanting to raise chickens, a growing interest in residential areas, might have to unite to change city ordinances!) Four largely individual and social obstacles, however, have to be overcome for large-scale home and community gardening to become a national backyard reality: 1) the mindset that neighborhood vegetable gardening (like individual recycling, home insulation, and private use of petroleum-saving cars) can make a difference in health, national energy independence, and global warming; 2) the willingness, at least at first, to be in the vanguard of a progressive behavior (especially because it may seem peculiar to most neighbors at first); 3) the modest learning curve required to acquire the know-how to become an effective gardener; and 4) the willingness to take the time and exert the energy to become physically able to do the simple gardening activities required to plant, cultivate, and harvest a garden.

Over time, I have come to believe that one's garden can, if labor intensive, be essentially gasoline free, leading to less energy demand, less pollution, and slower global warming. Although my experiences above have included first- and secondhand experience with rototillers, it has included more often spades, earth forks, and hand cultivators. The garden at the house we recently moved from, a raised-bed 20 x 20 feet (the largest allowed in the subdivision), was done completely by hand, my having given away my tiny, non-starting, gas rototiller to friends with a greater need and interest in repairing it. Although one might say that a rototiller does not use much gasoline and oil for a plot this size, it should also be pointed out that there is a petro-energy-based industrial process that manufactures the machine, the transportation of a borrowed one, and (minimalists might even argue) the space required to store it also required energy to build. About the time my rototiller broke down, I read Wm. S. Coperthwaite's *A Handmade Life*, a treatise on simplicity and ethics in design, beauty, work, education, and life. I began to see my small collection of power tools less as necessities, but more of ways to, as Coperthwaite says, "bull" my way through jobs[2], including tilling. I sweat more without the motorized tiller, yes, but I reduce my carbon footprint, reduce greenhouse gasses, reduce noise pollution, and increase my enjoyment, finding earthforking a therapeutic way to end a day at the office—and to raise my heart rate, build strength, and gain coordination to nuture plants in close quarters.

Since I began writing this chapter, my family has moved again, this time across town. In our previous subdivision, we were the only ones who I ever saw garden—except for one neighbor who grew herbs. Obviously, vegetable gardening is not considered a "normal" way to use yard space except by about 23% of Americans[3]. Our current 50 X 5 garden at the back edge of our new yard also seems to be the only one nearby. Almost everywhere I go, I cannot help but think that vegetable gardening is not generally considered to be something to do for practical, social, or philosophical reasons.

Social acceptance may be a high hurdle. Gardening is still considered—to use a term from the 1950s—nonconformist. Having one's own vegetable garden a dozen years into the 21st century may suggest to most poverty, eccentricity, or simply being old-fashioned—something out-of-step with big box stores, fast food, and fine dining. Yet, we know that many middle-class folks, as well as those with fewer means, grow at least some of their own. Vegetable gardeners are quite sane—perhaps more so for the responsible husbandry involved, and, given the realities of industrial food production, vegetable gardeners may, in fact, be very up-to-date in securing more colorful, tastier, more nutritious food.

Wanting to put a dent in environmental degradation and being willing to be the only gardener on the block are meaningless unless one acquires the knowledge, the know-how to make things grow. With reasonably fertile soil to begin, a water supply, and access to seed and a few tools, little knowledge is needed to get started on growing backyard vegetables. My active learning to grow green edibles began—with the encouragement of my friend's parents—with reading the backs of seed packets and the how-to articles that appeared in general-interest magazines in the 1970s—the same lessons that can be found today in any number of non-specialist magazines, ranging from *Mother Earth News* to *Popular Mechanics* and *Modern Frontiersman.* My involuntary childhood observation of neighbors growing rows of green cannot be discounted either, nor can the advice I have received over the years from experienced gardeners—people anyone can find by networking in their communities, going to farmers' markets, or, increasingly, becoming associated with a community garden where one rubs elbows with talented amateurs and volunteer professionals from schools and agricultural associations. But DIY vegetable gardening essentials, given one's interest, come from self-education, trial and error, and a little luck. I have found that simply drawing garden plans—showing what grows where—and keeping notes about successes and failures supplement my knowledge. Gardening essentials are quickly mastered and also help memory, spatial intelligence, and knowledge of basic life processes.

Perhaps most essential to successful season-long vegetable gardening is simply being able to get out and do it. I have seen many garden plots start great with enthusiastically prepared soil, neat seeding and planting, and clean initial weeding or mulching—only to collapse in mid- and late-season neglect. The usual reason given over a coffee or beer, after lack of time is lack of physical motivation. Often cited as well as a reason for not starting a garden at all, gardening is too often seen as simply *too much work.*

In a society focused so much on fitness, we are nevertheless one bogged down by what normally constitutes exercise—often in mechanized fitness clubs. I have nothing against the local gym, but home gardening presents the opportunity for often light to strenuous *fresh air* exercise with family and friends. As individuals look for reasons to exercise, for training partners, and the right facilities, one's backyard and nearby people offer a solid opportunity for improved fitness while growing good food.

I will not say that gardening provides an aerobic workout along the lines of running, swimming, or the many proprietary DVD workouts, but the advantages should be considered. First, gardening, by and large takes place outdoors. In a society that spends 90% of its time indoors in air of questionable quality[4], simply being out in the air is a big first step toward getting fit. Second, gardening provides a range of natural motion often working against gravity. Turning earth with a spade and fork, raking to smooth soil, bending at the waist planting and weeding, cultivating, pulling garden hose, wheelbarrowing tools and materials, and walking miles between garden and shed or garage during the growing season provide stretching, upper body lifting, and lower body conditioning. The key is not to do too much all at once at the beginning of the season or to lose commitment later. I used to wear myself out trying to prepare an entire garden plot in a day—turning soil, forking in peat and compost, raking smooth, staking out seed lines, and planting tomatoes and peppers. Since then, I have spread out these steps over several days and looked to the garden production and even aesthetics as incentive to continue the work over the growing and harvesting months.

Like the beginning of any physical regimen, starting gardening activity easily prevents soreness and the accompanying discouragement. Over the growing season, repetitious movement from replanting early growth vegetables, cultivating, turning compost, and harvesting in the fresh air all help to get one in better condition. Although gardening alone cannot result in wedding day weight, six-pack abs, and legs and arms of an Olympian, the gardener will at least have a daily foundation of activity sworn to by my 80-year-old gardening aunt—who, incidentally, probably could have fit into her wedding dress!

This last section offered a quick overview of answers to some common personal and social objections to exerting the power needed for home and community gardens. In my own state, it is often quoted that 90% of food is "imported" from other states and nations[5]. Growing even a small amount of produce means less food imported by truck, train, even jet, and fewer car trips to the supermarket. One gardener can reduce carbon footprint, many in a community reduce it much more. Overcoming any perceived stigma attached to gardening is increasingly dispelled by current environmental concerns and the environment of do-it-yourself and community gardening. Learning the simple procedures of home gardening can be mastered from published and community sources; likewise, regular and steady work in the garden results not only in a productive garden but also healthy time spent outdoors.

Of course, any gardener, consciously or not, is part of a movement that goes beyond her backyard and community garden. Today, a variety of perspectives directly and indirectly inform a new home and community garden ethic. The work of six individuals—though not all associated with small-scale food production--stands out for me. The first comes from someone known more for innovative housing and technological practices of traditional societies. Alluded to above, William Coperthwaite, a lifelong student of problem solving by first peoples, demonstrates the advantages of "a handmade life"—also the title of his book—that takes one in the direction away from power-assisted activity so much part of contemporary life. Although he does not address gardening directly, his focus on hand tools, human power, and ingenuity in moving toward the efficient and adaptable can all be applied to the home and community gardener.

Many years ago as an undergrad, I was privileged to meet Mr. Coperthwaite in one of his demonstration yurts. Impressed by his quiet hospitality, knowledge, and openness to my questions, I sent away for his do-it-yourself yurt instructions. Although I never built one (although I still own the hand-drawn illustrated plans), I contacted him many years later after reading his *A Handmade Life* that traced both his philosophy and some practical approaches to living today. Today, because of his book, I find a compelling argument to rely less powered tools and more on one's strength and skill. Although I have not given away all of my petro-tools and plug-ins—nor do I encourage others who otherwise use them for ecological good—with my recent move, I have sold my lawn tractor, given away my rototiller, and restricted my use of power-assistants, finding the earth fork and wheel barrow more than adequate and more satisfying in use.

Building on community power, Majora Carter, a pioneer of urban ecology, who I met a few years ago, shows how green can be created even in hardscapes. In the mission statement of her Sustainable South Bronx (SSB) foundation, her insistence on "environmental justice through innovative, economically sustainable projects that are informed by the needs of the community" ("Keynote")[6], have pointed the way to green roof gardens, new waterfront parkland, and ecological imperatives through the Bronx Environmental Stewardship Program (B.E.S.T.). Her example is timely, showing what is possible to encourage green living to rebuild even the most desolate urban landscapes. Given her achievements, would-be local gardeners in less-urban neighborhoods should take heart and understand how their efforts are made easier by having some private soil outside their back doors and realize how mulching, growing food, and learning about natural mechanisms of life are part of the much larger organized movement to make our earth more livable.

The reader may have detected tension between moving away from a more industrialized approach to local food production and greenspace—one involving machines such as rototillers, small tractors, and any number of mechanical-chemical devices—towards a more human-powered one involving muscle and brain power. William McDonough and Michael Braungart in "The Next Industrial Revolution" neatly separate the natural from the industrial in their discussion of waste streams. In their case for not mixing the byproducts of organic and technical processes, inspiration can be found for minimizing the industrial where the ultimate outcome is greening and, by extrapolation, minimizing the technical stream in gardening.[7] It may be that a large industrial component is necessary to *begin* gardening and general greening but this can be steadily reduced. At first, we may have to use tractors to break up sod or trucks (and their petroleum) to haul in soil for a raised bed garden. An initial investment in gardening tools, tomato cages, stakes, bean poles, lumber, and plastic film may seem appropriate to get started. Many continue to rely on industrial products to continue, but with smart buying of durable tools and garden accessories at the start, experience in the garden, and healthy living, many of the items that seemed so necessary at first will become less so. One may rediscover the satisfaction of using the spade, hoe, earth fork, and rake—all potentially lifetime tools. Heavy-duty tomato cages, bean poles, and garden stakes last many years. The lumber, if well-treated with natural preservative before installed as borders of a raised bed, will last years—and a stone one probably forever. The plastic film used to cover ground to keep down weeds can be replaced with straw purchased from a local farmer or with yard waste—all which can be later forked into the bed to enrich the soil. In brief, limiting our industrial garden inputs from the start can keep our gardens and planet healthier. By recognizing what is industrial and what is organic in our own small garden operations, we will rely only on the top quality manufactured items to begin and to use with each season of renewed growth and return to the soil. The result will be less long-term cost, fewer seasonal manufactured materials to dispose of, and more organic material to put back into the soil.

Barbara Kingsolver's approach in *Animal Vegetable Miracle* provides both information and spirit for green gardeners. Although many will not aspire to variety or amount of food her family did, we can appreciate her role as informant and cheerleader for serious, informed vegetable gardening. In particular, her effort to limit diet to what could be grown locally is a model step for reducing greenhouse gases. Involving her family and neighbors in her efforts provides an attractive social model, while recognizing the issues that sometimes arise. Finally, the satisfaction that results from the acquired knowledge, gardening itself, and resulting "miracle" of produce make vegetable gardening very attractive more than a fad or hobby.

In the last several years while I have been growing vegetables for family and friends, the community garden movement has been gaining momentum in my town. Several local experts have obtained permission to use public land, helped residents grow produce on plots and in wading pools at public housing facilities, applied for grants, partnered with the local food bank, and have taken the first steps in creating a local food co-op. The blend of expertise and energy from individuals in agricultural education, recreation, business, and community organizations has shown what committed individuals can accomplish to grow more local food. With each step, the word has spread: ad hoc neighborhood organizations have developed to expand gardens, safeguard crops, and engage more young people in the experience of growing food they eat. Of the many kinds of bridges from industrial to local food discussed, this last one leads me back to where I began with one important difference from the time of my childhood—the town kids today have guidance from many sources and the invitation to *participate.*

The seeds of new urban and suburban gardening have been dormant for a generation, yet their knowledge, skills, and necessity are near at hand and ready to be shared. Firsthand knowledge is still available, probably from more people than in many years: The old ones who have kept their plots active for two generations are eager to pass along what they know. Agricultural professionals understand the demand for their expertise in small scale, diverse operations on the scale of the backyard and community garden. Amateurs are feeling a greater sense of their ability to educate newcomers. Even small-plot, small-variety gardeners, such as this writer, have been approached by those who want to get started on their first gardens.

All of this first-hand knowledge is greatly supplemented by a large number of books and magazine articles found in libraries, magazine stands, and the web. Probably for the first time in 40 years and longer since young adults looked for societal solutions by going "back to the land," gardeners can build the bridge beneath their feet between large-scale monocrop agriculture and diverse small-scale gardening to meet the nutritional needs of people in all stations of life. Today, there is more knowledge and greater momentum seen in the late industrial era for food self-sufficiency.

Acquiring the skills required to garden are as easy to practice as briefly working alongside someone in the know or "connecting the dots" from a how-to guide, either short or elaborate. When I lived in Michigan, I heard a story about a homeless man who lived in an abandoned car in a devastated area of Detroit. Each morning when he woke up, he used a beat up hoe to scratch the debris and broken glass on the broken concrete around his car. One day amid these ruined buildings and vacant lots, the police on car patrol checked in on the old man. Driven by curiosity, the officer asked why the old man scraped around where nothing would grow. With a smile, the man said that if he didn't, his body would forget how to move. My dream is that this old man and many more disenfranchised have garden plots. Moreover, many of us do have land for gardens, tools waiting in our basements and garages, and the spare cash to buy seeds and a gardening book or two. With the desire to do so, we can get outdoors to grow more of our own food in an environmentally responsible way that will also build our neighborhoods and communities and attract those without land to build community gardens in their neighborhoods.

Ultimately, the combined knowledge, willingness, and diversity of individuals will bring the world closer to sustainable food independence, greater awareness of the human community, and true sustainability of world resources. The bridge between business-as-usual food production and local initiatives is becoming a reality—why not turn the soil and turn the tide?

Notes

1. Kingsolver, Barbara. 2007. *Animal Vegetable Miracle: A Year of Food Life*. New York: Harper, 2007.
2. Coperthwaite, William C. 2007. *A Handmade Life: In Search of Simplicity*. White River Junction, VT: Chelsea Green, 90.
3. National Gardening Association. 2009. "The Impact of Home and Community Gardening In America," http://www.gardenresearch.com/files/2009-Impact-of-Gardening-in-America-White-Paper.pdf.
4. William Bahnfleth, "Is Indoor Air Pollution Really a Problem?" *Research Penn State*, April 20, 2009. http://www.rps.psu.edu/probing/indoorair.html.
5. Daley, Bill. 2007. "Eating Close to Home: Eager to Embrace the 'Eat Local' Trend, Midwesterners Negotiate the Seasonal Challenges." *Chicago Tribune*, August 22. http://articles.chicagotribune.com/2007-08-22/entertainment/0708210758_1_number-of-farmers-markets-sustainable-food-local.
6. Sustainable South Bronx. 2007. "Majora Carter to Give WIU Environmental Summit Keynote Address." April 3. Quoted in the Western Illinois U. Home Page, http://www.wiu.edu/news/newsrelease.php?release_id=5298.
7. McDonough, William and Michael Braungart. 1998. "The Next Industrial Revolution." *The Atlantic*, October, http://www.theatlantic.com/william-mcdonough-and-michael-braungart/.

References

Bahnfleth, William. 2009. "Is Indoor Air Pollution Really a Problem?" *Research Penn State*, April 20, http://www.rps.psu.edu/probing/indoorair.html.

Coperthwaite, William C. 2007. *A Handmade Life: In Search of Simplicity*. White River Junction, VT: Chelsea Green.

Daley, Bill. 2007. "Eating Close to Home: Eager to Embrace the 'Eat Local' Trend, Midwesterners Negotiate the Seasonal Challenges." *Chicago Tribune*, August 22. http://articles.chicagotribune.com/2007-08-22/entertainment/0708210758_1_number-of-farmers-markets-sustainable-food-local.

"The Impact of Home and Community Gardening In America." 2009. National Gardening Association. http://www.gardenresearch.com/files/2009-Impact-of-Gardening-in-America-White-Paper.pdf.

Kingsolver, Barbara. 2007. *Animal Vegetable Miracle: A Year of Food Life*. New York: Harper.

McDonough, William and Michael Braungart. 1998. "The Next Industrial Revolution." *The Atlantic*, October. http://www.theatlantic.com/william-mcdonough-and-michael-braungart/.

Sustainable South Bronx. 2007. "Majora Carter to Give WIU Environmental Summit Keynote Address." April 3. Quoted in the Western Illinois U. Home Page. http://www.wiu.edu/news/newsrelease.php?release_id=5298.

Chapter 3: Where Does 'Local' Fit Under the Sustainability Tent? A Conversation with Dr. Joel B. Gruver

Joel B. Gruver and Thomas R. Sadler

Dr. Joel B. Gruver, Assistant Professor of Soil Science and Sustainable Agriculture at Western Illinois University (WIU), oversees activities at the Allison Farm, a certified organic research and demonstration farm in west-central Illinois. A specialist in soil science, Dr. Gruver investigates cover crops, crop rotation, mechanical weed control, organic soil amendments and other organic grain production strategies at the Allison Farm. He also teaches a full schedule of courses, including Introduction to Soil Science, Soil and Water Conservation, Nutrient Management, Soil Fertility and Plant Nutrition, Soil Properties, and Introduction to Sustainable Agriculture. Dr. Gruver is involved with many local food activities, including direct marketing of Allison Farm products, the Children's Tent at the Macomb farmers market, transplant distribution during the Heritage Day Parade, and the Macomb community garden. As a local authority on soil and organic growing techniques, Dr. Gruver has an important perspective on the current state of the local food movement in the region as well as future challenges. He provided written answers to questions concerning his background, education, and Allison Farm. In addition, on February 3, 2013, we sat down for a conversation about organic farming, local food networks, activism, trends in food production, and many other topics. His written and spoken answers follow.

As a child, did you have a garden or engage in other forms of local food production?

Our garden was a very important family activity. We spent so much time in the garden that my sister's first word was "pick." Soil preparation, including removing rocks using a large home-made sieve, planting, mulching, harvesting, composting, and canning were all major family activities.

How did you get interested in agriculture?

I started home gardening with my family as early as I can remember. My Dad had a rural heritage, was involved with 4H, and told us lots of stories about his experiences raising hogs including Baby Doll, his favorite adult female swine.

What is your academic background?

I have an undergraduate degree in Chemistry, from Principia College (1992). I hold a Master's Degree in Agronomy from the University of Maryland, College Park (1999) and a Doctor of Philosophy (PhD) degree in Soil Science, from North Carolina State University in Raleigh (2007).

What is the Allison Farm?

The Allison Farm is a 77 acre parcel of land that has been rented from private land owners (original Mr. Allison) since 1989. The land has never been farmed conventionally. The University's initial pesticide free agriculture project developed into a 100% certified organic farming program in 2009. The pesticide free agriculture program was launched by a former Agriculture department head Dr. Gerald Carlson who had been a USDA ARS (United States Department of Agriculture, Agricultural Research Service) research director before coming to WIU.

Do you bring your findings at Allison Farm into the classroom to share with students?

I show lots of pictures and some data as parts of lectures for most classes. I also take my lab classes on field trips to the Allison farm.

What are some examples of products that you've developed at Allison Farm?

Purple and gold popcorn is our main direct marketed product. We sell it in half pound zip-loc bags, 4 oz microwave bags and in bulk. We also sell wheat, rye, sunflower seeds and honey. There was little connection between the WIU campus and the Allison farm, prior to selling popcorn. Our purple and gold popcorn promotes school spirit and good will. It's not a big money maker because it is so labor intensive and is such a small part of the crops we grow.

The long-term vision is that we're developing a marketing concept that can be adopted by grain farms that would like to try something value-added. It would be great if lots of schools partnered with local farms to sell popcorn with the school colors, rather than candy. There are more than 20 K-12 schools in Illinois with purple and gold as their school colors. There are lots of possibilities since popcorn comes in many other colors including red, black, blue and white

Why do you have your students conduct interviews with people working in the agriculture industry?

The primary reason is for my students to gain current information (often information that is not available anywhere else) directly from people with specialized knowledge. The secondary reasons include: the development of student communication skills, the development of a culture of inquiry and openness to accessing primary sources and the compilation of useful bodies of knowledge to share with future students, interview candidates and the larger agriculture industry.

What is sustainable agriculture?

The simple definition is an approach to agriculture that actively attempts to address social, economic and environmental concerns. Another key element is flexibility/resiliency. Sustainable agriculture maintains the flexibility/resiliency to address unexpected challenges.

What are the challenges of sustainable agriculture?

One of the biggest challenges is the disconnect between consumers and farmers and the larger agricultural industry and the resulting distrust and lack of empathy. The dominance of the industrial/engineering paradigm in the development of agricultural technologies is a major contributor to many of the social, environmental and economic problems facing agriculture today. The industrial paradigm and its focus on production and efficiency is very valuable but differences between agriculture and other types of industrial production should not be ignored and the disciplines of ecology and sociology are needed to understand those differences.

What is the relationship between "organic" and sustainable?

At the heart of this relationship is the disconnect between consumers and producers. Most consumers have little knowledge about how agriculture is practiced today or personal connection to the farm community and thus have a challenge if they want to have an impact on how agriculture is practiced. Organic is the most well understood label that allows consumers to make their choices impact production. Organic is a concept that is legally defined and protected in this country by national standards. It allows farmers, food processors and vendors to differentiate their products in response to consumer concerns and capture more value. With knowledge, consumers have more confidence and vote with their dollars.

There's still a huge knowledge gap. Most consumers don't really understand organic regulations. Organic regulations in the U.S. are "process based rather than definitions of product quality such as nutrient content or the presence of pesticide residues." Organic regulations are more than prohibitions of synthetic inputs. Farmers must submit a whole farm plan documenting crop rotation and other conservation practices as well as all inputs and outputs to an independent certifier. The USDA has accredited over 50 certifiers but there are currently none based in Illinois. Certifiers contract with a 3rd party inspectors (i.e., not directly affiliated with the certifier) to evaluate whether the farms practices are in line with the farm's certification application.

The Allison farm is certified by the Midwest Organic Services Association (MOSA) based in Viroqua, Wisconsin, a certifier that has experience with research farms. This experience fosters greater flexibility with respect to the evaluation of novel products/practices at the Allison farm. The Allison farm has a complexity that is not normal for a corn and soybean farm. Each year we have 5-10 replicated experiments that require our fields to be divided up into dozens of plots that change from year to year. When a commercial farm becomes certified organic—a three year process— the farm is likely to keep the same fields in organic production. When evaluating whole farm plans, certifiers look for continuity of fields from year to year. We were advised to work with a flexible certifier that would understand our need for shifting plots and diversity of practices within fields.

Organic certification requires a paper trail of all inputs and outputs and transparency in record keeping. During an inspection, the inspector generally only asks the farmer to provide full documentation of select parts of their whole farm plan (e.g., tags from seed bags or clean truck affidavits). If the farmer is not able to provide the documentation, the inspector documents this and the certifier has to decide if the farm is engaged in fraud or simply has poor record keeping. Farms that will not benefit financially from certification are not likely to go through all the steps required to become organically certified. Many small farmers consider the process to be excessively onerous. For some people, it is not in their nature to keep good records. But there are benefits that come from the record keeping that are necessary to become certified. And in reality, the transparency in food production and processing that society is starting to demand requires good record keeping because most consumers don't have personal relationships with the producers of their food.

Is it cost effective to provide organic food on a large scale?

Large organic producers tend to be very entrepreneurial. They are generally more efficient and cost-conscious than small producers. Producing farm output cost effectively is a necessity and they take it very seriously.

Could we feed the country with organic food?

Organic farming is currently a small part of agriculture (<0.1 % of crop acres in IL are certified organic). As a result, current organic farmers could not produce the total amount of food consumed today and current conventional farmers could not seamlessly transition into organic. It would not be easy. But there are valid theoretical models that show it's possible. It would be difficult for organic farms to produce the tons of calories produced on conventional farms today. But we don't really need nearly the quantity of production we have today.

Side-by-side comparisons in industrial countries with high technology often show that organic yields are 10-20% lower than standard practices. In contrast, experiments in developing countries often show that organic yields are higher than standard practices. When you look at organic farmers and their conventional neighbors, you sometimes see large yield differences. Many organic farmers are not trying to produce equivalent yields to their neighbors. They may actually have greater profitability than their neighbors even with 50% lower yields.

Organic farmers often have different goals than their neighbors. If they are entrepreneurs, a primary goal is to have a successful business. But most organic farmers have other motivations including some that are philosophical. Why do farmers that self-identify as "organic" choose that identity? One hypothesis is they've had a paradigm shift, something happened: livestock died or someone was poisoned by pesticides. Social science research suggests that people who identify as organic often have a family culture of sustainability or conservation. There is something in their decision making that overrides economics and says this is what is moral to do.

The culture of agriculture involves an intertwining of lifestyle, family and business. Most farmers are willing to make great sacrifices to continue their farming lifestyle. In agriculture, low profitability tends to be more prevalent than high profitability. Farm families deal with extended low profitability in different ways. Family members often take another job. They may deliver the mail so they can continue to farm. Farming is a strong family tradition such that they would be a failure if they couldn't maintain what their family homesteaded and left to them as a legacy.

How does "local" fit under the "sustainability tent"?

Local addresses the disconnect between consumers and producers. Obviously if you have consumption of products near where they were produced, you have less disconnect. Normally if you have local consumption you have more of the consumer dollar flowing to the producer. You are addressing some of the issues such as the energy cost of agriculture. Producer decisions are more likely to focus on specific aspects of quality that are desired by consumers.

When there is no sense of who the consumer is, there is little definition of quality in a commodity. As soon as you have a concept of who the consumer is, you target the specific quality desired by the consumer. You have a social relationship. Farmers are valued by the local community and are seen as real people, friends.

Today Wal-Mart is the largest vendor of conventional and organic food in the US. At a store like Wal-Mart, the consumer experience for food and other home goods (e.g., hot dogs vs. paper towels) is not so different. The things that differentiate products at Wal-Mart are related mainly to price and convenience. Food packaging often includes romantic farm vistas but consumer awareness of current production methods has little bearing on consumer selection of products at Wal-Mart.

There are many interesting food system questions that students at WIU should be engaged in investigating. In Macomb, how much food is purchased at the farmers market versus at grocery stores? How do people make purchasing decisions? Where do students and professors buy food? How much of the food is consumed in home or out of home? It's valuable to investigate how the current food system operates. It's just as important to understand how we currently keep the shelves of Wal-Mart full as it is to understand alterative food systems. We have a need for continuous re-supply that requires a transportation infrastructure that seems reliable but could break down.

We need to be realistic about the efficiencies that have been achieved. One ton of food can be moved 500 miles with one gallon of diesel fuel on a train. Moving things by barge is even more efficient. The energy efficiency of the vehicles bringing food to the Macomb Farmer's market is low, compared to a semi-truck or freight train bringing products from California. But even if a semi-truck is one hundred times more efficient than a small farm truck, less energy may be required to transport food from a local farm. Students should analyze these scenarios rather than assume that it's always more energy efficient to eat locally.

Do we need metrics to judge a local food network?

To achieve an efficient local food network, we should have metrics that we apply to our local food system that we are willing to accept: simple metrics that identify success as well as opportunities for improvement. We need to be teaching about these types of metrics at the University and helping consumers to understand them. Some people grow up with a culture of conservation but this is not the norm in our society today. For people with a culture of conservation, there are things that as a family you talk about that perpetuate the culture: you think about where water comes from, energy efficiency and over-packaging. Simple metrics can help people who did not grow up with a culture of conservation to understand the value of local food.

How would you characterize the local food movement in this region?

I think we have an active farmers market. It would be valuable to know from the vendors what portion of their output they are growing. Also, what portion of income comes from the farmers market? Are the sellers farmers? These are numbers that would be useful to help us understand the local food economy. I wonder what the total sales are on a Saturday farmers market. My guess is the top vendors are selling more than $1000. Most successful vendors around the country are selling more than $3000 per day; I don't think we're close to that. As a college town we have a higher percentage of consumers that are interested in the origin of our food. We've made progress. The local food economy has strengthened. More people are coming.

The Food Initiatives Group (FIG) helped with promotion and the children's tent. But I don't know how much of an impact it had. The food co-op development that's going on is proof that there is a significant contingent of people who are interested in the alternative food system. They have 225 members, but need 400 to take the next step. Going from 225 to 400 will be difficult. The last 20 have come through more effort. The next 180, I don't know how they'll get them.

The food co-op is an example that came out of FIG. The same thing with the community garden. FIG was successful and still has a role to play. I don't see FIG as gone; it's dormant. There are certain elements of FIG that continue: transplant distribution, the children's tent. In terms of the community garden, I'm more intimately involved than the food co-op. The Macomb Community garden is a work in progress but we are off to a good start. We have a fantastic site with amazing soil but need to grow to include more people and to solve some of the social sustainability riddles.

What other statistics do we need?

It would be great to have students survey people at the Macomb Farmers market. How much did you spend at the farmers market? Why do you come? Do you bring children? Do you buy certain things in season only at the farmers market? There is this culinary culture that is to some extent antithetical with local food where people like to watch the cooking channel, read magazines, identify recipes, and decide what to cook before they get ingredients. The culture most conducive for local food is a culture of going to the farmers markets and then deciding what to cook.

Do we need a different culture of cooking?

We need a different culture of cooking that would make the farmers' market style of buying a better fit with our eating habits.

How do we establish a different culture of cooking?

In Macomb, FIG has organized two different types of local food meals: two fancy meals, where people purchased a ticket in advance and were served at least five courses and one free buffet style meal. For the fancy meals, we could have charged a lot more money. With the free meal, I hope people left thinking "I could cook that at my house."

It's easier to have a fancy meal and get foodies to pay $30 a ticket than to get low-income people to come for a free meal. It would be valuable to have more accessible food for everyone. The community garden can serve that purpose.

In our region's future, how do envision the local food movement?

I don't think that we have saturated the higher income demographic. There's still room for growth in getting those people involved. We have more progress to make in the low-income demographic. Our local institutions should be encouraged to support local agriculture through their buying decisions. . Very soon we are going to have a meeting to discuss potential gardening activities at the McDonough District Hospital. Who knows what will come out of it: flowers for patients, horticultural therapy, or even food for the hospital? Lots of institutions in Macomb could participate more fully.

What I was working on this morning for my water conservation class is the current state of natural resources in Illinois. We have two types of landowners, public and private: 96.9% of Illinois is owned privately, but within this private ownership there are opportunities for stewarding natural resources. Also, every elementary school should have both wildlife habitat and a garden. If all universities, community colleges, and secondary school systems had habitat and gardens on campus we would have better education and probably better food. In terms of habitat, there is great value in connectivity. For habitat to work, organisms must be able to go from place to place. With local food, there is great value in the cross pollination of ideas between educators and students.

Chapter 4: More than Just Food: Wisdom, Vitality and the Local Food Movement

Thomas R. Sadler

Introduction

Contemporary economic and environmental challenges such as drought, peak oil, climate change and soil depletion alter economies and their supporting food networks. As one example, the drought of 2012, the worst in the United States in fifty years, drove up the price of milk, beef and pork, with lower corn and soybean yields increasing the price of industrial food (Lowrey and Nixon, 2012).

Because industrial agriculture requires a continuous flow of fossil fuels and global networks of distribution and leads to health-related and environmental externalities, many academics and activists argue for the strengthening of alternative food networks, especially those that entail smaller-scale and local farming (Schramski *et al.*, 2011 and McKibben, 2007). Local food networks reduce fossil fuel consumption, enhance air and soil quality and may stimulate community development. These energy, environmental and economic benefits serve as the basis of the local food movement (Hewitt, 2009).

But James E. McWilliams, in his book *Just Food: Where Locavores Get It Wrong and How We Can Truly Eat Responsibly* (2009), argues against the development of local food networks. His premise is that a focus on local food networks masks other steps in the production and distribution processes that harm the environment, such as pesticide application, fertilizer outlay and energy inefficiency. Moreover, with respect to meat, small-scale production leads to environmental problems, including more methane from grass-fed cows, more land for grazing, and extensive waste from processing cows before the end of their natural lives.

However, McWilliams' focus on the appropriate system that would feed the ten billion people on earth at century's end is a different question than how communities may enhance their food security with a local food network. The reality is that local food networks are growing. According to Low and Vogel (2011), in a report for the U.S. Department of Agriculture, revenue from the sale of local food has been radically undercounted, exceeding $7 billion in 2008.

Since 2004, in the U.S., the number of farmers' markets has nearly doubled (Strauss, 2012). According to O'Hara (2011), over 4000 Community-Supported Agriculture (CSA) programs exist. This evidence reveals a number of benefits of local food networks: health, knowledge, energy savings and community activism (Martinez *et al.*, 2010).

As with many food-related challenges, Wendell Berry explains that change will come from people in communities doing things differently:

> No great feat is going to happen to change all this; you're going to have to humble yourself to be willing to do it one little bit at a time. You can't make people do this. What you have to do is notice that they're already doing it. (Bittman, 2012)

By elucidating the intricacies of a dynamic local food network, this chapter demonstrates that McWilliams' objection to local food stems from a bias toward big business. More specifically this chapter argues that, while a local food network must reform the abuses of industrial agriculture and preserve the profit motive, the values of wisdom, community vitality, and economic opportunity—so important for a sustainable future—*require* responsible and healthy food choices. This chapter makes clear that change occurs—as the Wendell Barry quote reveals—when community members demand (and are willing and able to pay for) food items that will contribute to and not take away from the goal of community vitality.

In this process, local farmers, households and businesses must all serve as stakeholders. To consider an example, the objection of McWilliams (2009) that grass-grazing cows require more land than their grain-fed counterparts is true. But a local food network encourages farmers to transform their land, using fewer acres for the production of commodity crops, and more acres for meat and produce. When compared to the industrial alternative, this transformation requires fewer fossil fuels. In terms of an overall contribution to a community, more land allocated to grass-grazing cows is more than offset by the application of less chemical fertilizers and pesticides, fewer fossil fuels, higher soil fertility and other net benefits.

The context of this chapter is the author's hometown of Macomb, Illinois and the region of central Illinois. This area currently lacks a strong local food tradition, even though Macomb resides in the country's agricultural heartland. But the area boasts several elements of a growing local food network: a thriving CSA; a local Food Initiatives Group (FIG); the Illinois Stewardship Alliance in nearby Springfield; a community garden; and a network of people who are working to establish a local food co-op.

This chapter demonstrates four points. First, a strong local food network establishes values of wisdom, collective sustainability and economic vitality. Second, when local food producers sell their output in the widest variety of markets—including direct-to-consumer outlets and intermediate channels—the producers have the greatest potential to achieve economic vitality. Third, food network goals include a growing local food economy, economic opportunity and positive environmental and social outcomes. Fourth, this goal-oriented framework helps to determine how effectively a community addresses the problems of economic, environmental and food insecurity.

Values of a Dynamic Local Food Network

The narrow focus of neo-classical economics on self-interest, profit maximization and the assertion that "more is better" creates conflict over natural resources between humans, societies, species and generations. In recent years, with the financial crisis (2007 – 2008), higher energy prices (2008 – 2012), and the spike in global food prices (2008, 2012), the world has experienced debt-fuelled, fossil-fuelled and consumption-based problems.

A myopic focus on economic growth and self-interest is not suitable for the development of a local food network. The reason is that this narrow vision ignores the reality that members of a dynamic local food network contribute to the welfare of others. Food producers assume economic and environmental responsibility for themselves *and* their customers. Merchants, who contribute value-added products, help to create producer *and* consumer surplus. Consumers purchase and cook healthy food for their families, but *also* encourage local food production.

A local food network therefore contributes to the process of human development by expanding people's food choices, building stronger communities, and strengthening the natural resource base. Such a network enhances the level of interconnectedness between local food producers and consumers. This "circular flow" between food producers and consumers, so important for a local food network, reflects values not of neo-classical economics, but of a wisdom economy. While questions of growth and development are relevant and addressed in this chapter, the values important to a local food network are wisdom, collective sustainability and economic vitality.

In an article in *Ecological Economics*, Murtaza (2011) establishes a link between these values and economic structures. As society's knowledge increases, values and economic structures co-evolve. Initially, an economic network is driven by its leaders. Over time, the network influences larger sections of the population, who in turn alter social structures to provide focus and motivation for their lives.

The extent to which social structures create positive and lasting change depends on how a society establishes its motivations. One possible motivation, profit maximization, leads to income inequality and economic instability. While twenty percent of the world's population has achieved a relatively high level of wealth, almost forty percent is moderately or extremely poor (Sachs, 2005). But a different focus on wisdom—inherent in local food networks—inspires "unbounded" self-actualization, leading to the pursuit of the greatest social good (Murtaza, 2011).

This dimension of thinking does not separate collective sustainability—the maintenance of the natural resource base—and personal welfare. When the goal is community vitality, a focus on the underlying value of wisdom—and not unbridled self-interest—may contribute most effectively to the development and expansion of the local food network (Murtaza, 2011). In this context, not only does the pursuit of personal welfare require belonging to an "interrelated and interdependent" humanity, but ecological sustainability requires a sense of responsibility for others (Bina and Vaz, 2011).

The viability of a local food network depends on the willingness of community members to participate. Ultimately, households that purchase food items in "conventional" market outlets, such as grocery stores or Wal-Mart, must consider the option of buying from local food producers in different outlets and channels. These same community members must then learn to cook with locally-sourced ingredients. The goals of human health and ecological integrity influence this behavior, but also the importance of a strong local food network *as an end in itself.*

This focus on wisdom and interconnection is not to say that economic principles are unimportant. On the contrary, a strong local food network requires local food producers to earn economic profit. But they must also serve as *active agents*, acknowledging that ecological responsibility accompanies market freedom. A narrow focus on profit maximization precludes both the establishment of an "ethic of character," which includes ecological considerations, and development of a local food network, which strengthens the connection between local food producers and consumers (Bina and Vaz, 2011).

The objection of McWilliams (2009) to the establishment of strong local food networks is relevant here. McWilliams is partially right that "scaling down food production eating local fare is not in and of itself going to feed the impending 10 billion in a sustainable way." But an emphasis on local networks does not mean "scaling down" food production. On the contrary, a market-driven network, if successful, provides the output that consumers demand. A movement toward greater consumption of locally-sourced food may accompany an increase in overall food production. While this chapter rejects McWilliams' (2009) arguments for genetically modified crops, chemical treatment of soil, and a preference for large agricultural operations, the development of regional "foodsheds" is an idea worth exploring. Having examined the values important for dynamic local food networks—wisdom, collective sustainability and economic vitality—the next section discusses the model of a local food network.

Model of a Dynamic Local Food Network

A local food network requires each of the following: food producers (farmers, gardeners and food merchants), markets for the sale of local food, and evolving conceptions of what food to buy and how to eat. Largely decentralized and dependent on activist citizens to create a marketplace, a local food network may serve as a component of community development.

The supply side of a strong local food network includes small-scale and medium-scale farmers that sell food in direct-to-consumer outlets and intermediate channels (Table 1). Examples of direct-to-consumer outlets include farmers markets, roadside stands, farm stores, CSAs, food co-ops and food trucks. Examples of intermediate channels include sales to grocers, restaurants and regional distributors. With intermediate channels, after markups from grocers and wholesalers, consumers pay about four times what a farmer receives (Strauss, 2012). With direct-to-consumer outlets, local food producers retain most, if not all, of the revenue from the sale of their produce. On a per-unit basis, local producers may receive up to seven times greater net revenue from selling locally than in industrial markets (O'Hara, 2011).

Direct-to-consumer outlets	Intermediate channels
Farmers markets	Grocers
Roadside stands	Restaurants
Farm stores	Regional distributors
Community-supported agriculture	
Food co-ops	
Food trucks	

Table 1: Local food outlets and channels

While many communities have a natural disposition for locally-sourced food—including Hardwick, Vermont, Fairfield, Iowa, Madison, Wisconsin and many others—our community of Macomb, Illinois does not. Macomb lacks the institutions, market processes and wide-spread attitudes necessary for a dynamic local food network.

Nevertheless, as this and other chapters in the book explain, these components of a strong local food network are evolving, even starting to thrive. More local food producers and consumers exchange food, money and ideas about planting, harvesting, cooking and eating. As Ben Hewitt writes in *The Town that Food Saved* (2009), "The redrawing of our straight-line agriculture network into something even vaguely circular is a generational and perhaps multigenerational task." This time element demonstrates that local food producers, consumers and markets develop in tandem.

No universally accepted or legal definition exists for "local" food. The term may refer to geographic characteristics, particular techniques of distribution or market qualities. Urban residents may consider as local the food that comes from surrounding counties. In contrast, rural residents may consider food from their neighbors as local. The so-called "100 mile diet" that gained popularity in recent years chooses an arbitrary numerical radius.

Social embeddedness—the connections between producers and consumers often synonymous with local food networks—may serve as the essence of the term local. In this context, a "short" food supply chain facilitates a sense of connection: by providing a clear signal with respect to the origin of food, consumers possess knowledge concerning the farmer's methods of production, use of natural inputs and markets for exchange. Variables including economies of scale, distribution, and seasonality serve as important market characteristics and differ when evaluating direct-to-consumer and intermediate channels (Martinez *et al.*, 2010).

Another important and evolving characteristic of local food networks concerns the method of financing: producers require financing before their output is available. With industrial agriculture, early-season financing from banks, subsidies from the federal government, and crop insurance facilitates an efficient marketplace, insuring against large-scale losses from unforeseen weather patterns or environmental disasters. For the small-scale farmer, however, obtaining the resources necessary to plant and then harvest proves challenging. The CSA model, in which members provide their payment in the spring so the farmer may purchase seeds and begin planting, serves as one economic framework

(McIlvaine-Newsad *et al.*, 2004). A local food co-op, in which members contribute money equal to the value of shares and promise to spend a certain amount at the co-op each month, serves as another (Hewitt, 2009). An additional method of financing, the "slow money" movement, links investors to the places they live, creating new sources of capital for small food producers (Tasch, 2010).

A final characteristic of a local food network serves to close the loop between producer and consumer: food activism. In any marketplace, consumers "vote" with their dollars. As households spend a greater percentage of their budget on locally-sourced food, the marketplace provides producers with the incentive to maintain or expand their range of output. But food activism, when citizen groups strengthen local food networks or block the expansion of deleterious outcomes of industrial agriculture, advances the process. When activist groups establish lasting local institutions that influence public opinion, they form a framework for evolving local food networks. In his influential book on environmental activism and social justice, *Blessed Unrest* (2008), Paul Hawken argues that these activities are necessary to "save the entire sacred, cellular basis of existence." Activism helps to create the local food networks that provide an alternative to industrial food, but activism also provides feedback to local food producers: their actions are important for the future of the community.

In central Illinois, two recent examples of food activism demonstrate this point. The first example of food activism, Rural Residents for Responsible Agriculture, formed in 2011 to prevent the establishment of a confined animal feeding operation (CAFO). The owner of the proposed hog farm wanted to produce more than 200,000 piglets per year. A record of pollution by the company plus the desire of many rural residents to prevent the invasion of their properties by the feeding operation galvanized the group to voice their disapproval. Because of this and other opposition, in 2012 the CAFO's application expired without approval by the Illinois Department of Agriculture. The result was a victory for Rural Residents for Responsible Agriculture and the community. But this activism also highlighted the key problem of industrial meat production: ecological degradation. Learning more about the industrial process through public hearings and the press, many citizens bought meat from local suppliers, strengthening the local food network.

The other example, the Macomb Food Initiatives Group (FIG), established in the fall of 2007, galvanized many community members to become involved in the local food movement. The FIG encourages the community to support a fair, sustainable and local food network. While the goals of the group are to educate consumers about food choices, engage consumers in preparing food, foster food-related networks, and advocate policy changes, the group brings together diverse community members. Programs include community dinners, resources for the establishment of a community garden, and an ongoing effort to create a local food co-op. While two of the local food dinners were ticketed events and sold over one hundred dinners each, another event, a soup dinner, was free to the public. With funds generated from FIG memberships and fundraising, the soup dinner operated in the context of the monthly free meal that community churches oversee (Figure 1).

Figure 1: Community dinner with locally-sourced food

Goals of a Dynamic Local Food Network

Three goals of a dynamic local food network establish a framework for evaluation. The first goal—growth of the demand and supply sides of the market—recognizes the importance of the outlets and channels in Table 1. If a local food network is to thrive, not only must these outlets and channels provide an appealing and cost-effective alternative to traditional grocery stores and Wal-Mart, but the town's citizenry must choose to participate in these market options. That is, as the supply of local food increases, so must demand. Information about the benefits and costs of the outlets and channels empowers citizens to consider all consumption alternatives, including the possibility of buying directly from a local food grower or merchant. The creation of a healthy local food network provides an ideal way to leverage that capacity to support local food producers.

The second goal is to create economic opportunity in the form of jobs. The extent to which farmers and merchants produce, process and distribute locally-sourced food items, generate income, and employ workers depends on both local and global markets. In particular, the potential to create jobs with the local food network depends on the value of the community spending multiplier. The multiplier is the number of times a dollar cycles through an economy. The multiplier serves as a proxy for the economic power of local spending. In terms of a local food network, the higher the multiplier, the more a typical dollar cycles through the system.

When farmers sell directly to consumers in local outlets, they keep a much higher percentage of each dollar of sales, when compared to intermediate channels and global markets. The farmers then spend the income in the community. This market activity increases the community spending multiplier.

When communities establish markets that link local producers and consumers, small farms benefit.

The more that community dollars flow to local farmers and business owners, who then spend the same dollars locally, the higher the "local productive economic activity" of the community. As an example, Illinois residents spend $48 billion annually on food (Illinois Local Food Farm and Task Force, 2009). But very few of these dollars are spent on locally-sourced food in the state. Keeping a small percentage of this spending local would translate into many food-related jobs. The development of a local supply chain that shortens the distance between farmers and households requires greater local food infrastructure, awareness and access. Communities, not state governments, are in better position to establish these important components of local food networks.

The third goal of a local food network—to improve environmental and social conditions—depends fundamentally on the ability of food producers and community members to develop the institutions and attitudes necessary for a local food network to thrive. Recent research on the topic focuses on "econetwork" services to and from agriculture (Stallman, 2011; Swinton et al., 2007; Zhang *et al.*, 2007). While these studies are important and valuable to this discussion, a local food network may create additional benefits to a community or region, in addition to econetwork services. Local food production, for example, may improve air, water and soil quality, but local food networks may also contribute to community development. The next section considers these environmental and social conditions.

Environmental and Social Conditions

Adapted from Stallman (2011), Swinton et al. (2007) and Zhang *et al.* (2007), Table 2 lists environmental and social conditions and descriptions that result from local food networks. Imagine these environmental and social conditions along a continuum (Stallman, 2011). For example, an important regulating service of local food production is soil retention. At one end of the continuum, a high level of soil retention exists. The other end represents soil erosion. As another example, participation in the process of food production may contribute to higher levels of spiritual well-being at one end of a continuum, but a loss of well-being at the other.

Understanding these outcomes of local food networks allows for an evaluation of the system. If environmental and social conditions are closer to the disservice end of the continuum, the network is not providing benefits. As social and environmental conditions move to the positive end of the spectrum, the local food network strengthens the community or region.

Condition	Description
Econetwork services (ES)	Natural capital which provides vital goods and services to humans
Production services	Food, fiber and fuel
Regulating services	Soil retention, pest control, pollination, water purification, habitat

Supporting services	Nutrient recycling, water filtration, biodiversity, soil fertility
Cultural services (CS)	Services that enhance culture and human well-being
Agricultural lifestyles	Family and individual opportunity for quality of life
Esthetic landscapes	Potential for increasing value and appreciation for landscape
Spiritual well-being	Connection between access to the land and individual well-being
Externality elimination (EE)	Elimination of spillover costs from industrial agriculture
Air quality	Elimination of chemical emissions into the air
Water quality	Elimination of chemical emissions into the water
Soil quality	Elimination of chemical emissions into the soil
Community development (CD)	Networks of problem solving, community building, interaction
Democracy	Community members solving problems of food production
Institutions	Creation of demand/supply institutions for the local food network
Policy	Government policy that facilitates growth of local food network

Table 2: Environmental and social conditions
Sources: Stallman (2011), Swinton *et al.* (2007), Zhang *et al.* (2007)

Humans manage agricultural econetworks to optimize the production of food and fiber. The econetwork services from food production depend in turn on a network of supporting inputs, including soil fertility, irrigation and sunlight. Food production may also receive dis-services—such as droughts and undesired species—that increase cost or reduce productivity. The ability of food providers to minimize the flow of dis-services depends on how agricultural processes are managed and the diversity of the landscape and natural econetworks (Zhang *et al.*, 2007).

Cultural services, including agricultural lifestyles, esthetic landscapes and spiritual well-being, contribute to a community or region. The ability of farming families to establish agricultural lifestyles provides economic opportunity. An esthetic landscape that combines annual food production, perennial trees and plants, and forest and prairie econetworks enhances the rural value of place. The ability to connect with the land and participate in the local food network may serve as an element of spiritual well-being, the multifaceted ability to be content with place, opportunity and relationships (Berry, 1996).

Externalities—costs that accrue to individuals not involved in an economic process or transaction—exist when producers externalize certain aspects of production on society. Examples include polluting emissions into the air or chemical waste into the water or soil. In the presence of high costs of negotiation between polluters and victims, society suffers from the reality of negative externalities. With public policy or a commitment approach from individual producers, the reduction of negative externalities occurs (Tomer and Sadler, 2007). Local food networks, in which consumers possess a greater understanding of the production process of farmers, have the potential to grow without creating the negative externalities associated with fossil-fuels and environmental degradation.

Community development seeks to empower individuals, so they may obtain the skills, knowledge and social capital necessary to improve community affairs. In this context, a vision of a strong local food network represents a patient investment by community members in shaping their future. Local food dollars multiply throughout the local economy, weaving a "community web." Local food networks respond to community needs, bring more voices to the marketplace, and mobilize community members to serve as informed food producers and consumers. Because CSAs, food co-ops, farmers markets, local food dinners and other food initiatives have their roots in community-based resources and local ownership, they establish far-reaching relationships and create jobs, two fundamental elements of community development.

The Case of West Central Illinois

The examples of local food markets in Table 1, the three goals, and the outcomes in Table 2 may be used to assess local food networks in communities or regions. If a local food network lacks certain outlets and channels in Table 1 or does not achieve the three goals, for example, community members may create additional institutions such as food co-ops or CSAs or focus on the outcomes in Table 2.

In Macomb, Illinois the Barefoot Gardens CSA offers annual memberships, which provide weekly baskets of seasonal produce, from May until December. At Barefoot Gardens, the proprietor, John Curtis (a co-author of this book), raises a variety of vegetables, fruit and flowers. A community institution since 2003, Barefoot Gardens offers local and organic food options. But Barefoot Gardens also serves as a "third place," neither work nor home, where people gather to discuss community challenges. While the main purpose of the CSA is to provide healthy and delicious food to members, the beautiful CSA gardens serve as a classroom for individuals learning about gardening and food preparation.

In 2010 Macomb residents and Western Illinois University faculty members Joel Gruver, Heather McIlvaine-Newsad and Rob Porter (all contributors to this book) created a community garden. Providing access to fresh produce for all interested members of the community—including individuals and household that may lack the financial means to buy a share in a CSA—the community garden encourages group planting and harvesting, so gardeners interact with and learn from others. While the goal of the community garden is to provide a setting for the growing of local food, it also brings people together to share growing tips, cooking ideas and friendship.

A local food co-op movement began with early discussions among FIG members, during 2007 and 2008. A plan of starting small with an offering of local produce in a downtown space has grown into an idea of creating a full-fledged co-op grocery store, similar to other food co-ops in the region. Of all the local food initiatives in the community, this co-op has the greatest potential for job creation. When fully functional, co-op members will operate the store, order the food items and organize seminars on cooking and food preservation.

Figure 2: Barefoot Gardens CSA

The trend in local food networks is a movement toward greater integration. But job creation does not depend exclusively on the development of the local food market. Especially with respect to value-added forms of output, access to national and global markets helps to strengthen the market position of individual merchants. As one example, Joel Gruver, working with students from Western Illinois University, uses corn from Allison Farm—the university's organic farm—to make organic popcorn, packaged and sold throughout the region. This value-added product not only brings recognition and revenue, but reaffirms the idea that sustainable farming has a place in the economy.

Conclusion

Considering all outlets and channels, the local food economy in West Central Illinois is small but growing. While not yet achieving the goal of an engine for job creation, the local food network has led to some of the tangible outcomes in Table 2, especially cultural services, community integration, and even more backyard gardening – including this author's garden (Figure 3).

An increasing demand for local food allows some farm families to sell to local markets and continue living the agricultural lifestyle. An improvement in the aesthetic landscape from diverse local food production is apparent. Aspects of community development continue to evolve, including local food activism and the growth of local food institutions. The local food dinners bring together households from different socio-economic backgrounds. The Food Initiatives Group raises awareness about locally-sourced food options and CSAs continue to provide local produce.

Figure 3: Thomas R. Sadler's backyard garden

In the heartland, the local food network in central Illinois is ready for growth. In this region, growth in the local food network is about more than just food. Growth in the local food network may help to improve environmental quality, offer healthier food choices, provide a much-needed boost to the local economy, and serve as a way for community members to come together to improve their well-being.

References

Berry, Wendell. 1996. *The Unsetting of America: Culture & Agriculture*, San Francisco: Sierra Club Books.

Bina, O., and S.G. Vaz. 2011. Humans, environment and economics: From vicious relationships to virtuous responsibility. *Ecological Economics* 72:170 – 178.

Bittman, Mark. 2012. Wendell Berry, American Hero. *The New York Times*, 25 April.

Hewitt, Ben. 2009. *The Town that Food Saved*, New York: Rodale Books.

Johnson, Kirk. 2012. Small Farmers Creating a New Business Model as Agriculture Goes Local. *The New York Times*, 1 July.

Low, S.A., and S. Vogel. 2011. *Direct and Intermediate Marketing of Local Foods in the United States*, Washington DC: U.S. Department of Agriculture.

Lowrey, Annie and R. Nixon. 2012. Severe Drought Seen as Driving Cost of Food Up. *The New York Times*, 25 July.

Martinez, S., M. Hand, M. Da Pra, S. Pollack, K. Ralston, T. Smith, S. Vogel, S. Clark, L. Lohr, S. Low, and C. Newman, 2010. *Local Food Networks:*

Concepts, Impacts, and Issues, Washington DC: U.S. Department of Agriculture.

McIlvaine-Newsad, H., C.D. Merrett, and P. McLaughlin. 2004. Direct from Farm to Table: Community Supported Agriculture in Western Illinois, *Culture and Agriculture* 26:149-163.

McKibben, Bill. 2007. *Deep Economy: The Wealth of Communities and the Durable Future*, New York: Henry Holt and Company.

McWilliams, James E. 2009. *Just Food: Where Locavores Get It Wrong and How We Can Truly Eat Responsibly*, New York: Little, Brown and Company.

Murtaza, Niaz. 2011. Pursuing self-interest or self-actualization? From capitalism to a steady-state wisdom economy, *Ecological Economics* 72:577 – 584.

Lowrey, A. and R. Nixon. 2012. Severe Drought Seen as Driving Cost of Food Up. *The New York Times*, 25 July.

Norberg-Hodge, H., T. Merrifield and S. Gorelick. 2002. *Bringing the Food Economy Home: Local Alternatives to Global Agribusiness*, Zed Books.

O'Hara, Jeffrey K. 2011. *Market Forces: Creating Jobs Through Public Investment in Local and Regional Food Networks*, Union of Concerned Scientists.

Pollan, Michael. 2007. *The Omnivore's Dilemma: A Natural History of Four Meals*, Penguin.

Sachs, Jeffrey D. 2005. *The end of poverty: economic possibilities for our time*. New York: Penguin Press.

Schramski, J.R., Z. J. Rutz, D.K. Gattie, and K. Li. 2011. Trophically balanced sustainable agriculture, *Ecological Economics* 72:88 – 96.

Stallman, Heidi R. 2011. Econetwork services in agriculture: Determining suitability for provision by collective management, *Ecological Economics* 71:131 – 139.

Strauss, Robert. 2012. This Little Farmer Went to Market. *Bloomberg Businessweek*, 21 May.

Swinton, S.M., F. Lupi, P.G. Robertson, and S.K. Hamilton. 2007. Econetwork services and agriculture: cultivating agricultural econetworks for diverse benefits, *Ecological Economics* 64:245 – 252.

Tasch, Woody. 2010. *Inquiries Into the Nature of Slow Money: Investing as if Food, Farms, and Fertility Mattered*, White River Jct.: Chelsea Green Publishing.

The Illinois Local and Organic Food and Farm Task Force. 2009. *Local Food, Farms & Jobs: Growing the Illinois Economy*, Springfield: Illinois General Assembly.

Tomer, J.F. and T.R. Sadler. 2007. Why We Need a Commitment Approach to Environmental Policy, *Ecological Economics* 62:627 – 636.

Zhang, W., T.H. Ricketts, C. Kremen, K. Carney, and S.M. Swinton. 2007. Econetwork services and dis-services to agriculture, *Ecological Economics*, 64:253 – 260.

Chapter 5: A Sense of Place

John Curtis

Any thinking or writing on local systems must begin with an actual place in mind. My place, and the place of many of the contributing authors in this book, is in and around a small town called Macomb, in west central Illinois. Situated in the center of McDonough County, Macomb is a relatively isolated town of almost 20,000 people. The closest major city is St. Louis, three hours distant, with smaller cities at least an hour away. In 2013, Macomb is like many other towns of its approximate size in the Midwest. It has a historic courthouse and a faded town square. As in most towns of any size in this region, most retail business is done on the edges of town where the big box stores and fast food chains are concentrated.

Macomb is the home of Western Illinois University. When school is in session, students make up fully one half the population of the town. Needless to say, the university has had a huge impact on the community – not the least of which is that it has kept Macomb alive during some tough economic times.

The rest of McDonough County has not fared as well as Macomb. A tour of our rural landscape tells a story of agricultural industrialization and rural decline. Commodity corn and soybean production dominates most of our region. During the summer in many parts of the county, these two crops stretch in huge expanses as far as the eye can see. The rest of the year, the soil is bare and the landscape is empty; a great desert that stretches, largely unbroken, all the way from Macomb to the suburbs of Chicago almost four hours away.

Industrial scale animal operations are another feature of our area. The long, metal buildings typical of modern hog confinements seem to increase in number every year. Except for the occasional cattle feedlot or pasture, it is unusual to see animals on the landscape. It is also unusual to see a human, unless it is an agribusiness man or (more likely) a hireling climbing out of one machine and into another. Sighting two or more rural people, together, out of doors, is rare indeed.

Though it is by far the largest and most prosperous, Macomb isn't the only town in McDonough County. In fact, the 2012 Plat Book for the county lists twenty-one other small towns that have their own zip codes. These include Adair, Avon, Bardolph, Blandinsville, Bushnell, Colchester, Colmar, Good Hope, Industry, LaHarpe, Littleton, Marietta, Plymouth, Prairie City, Raritan, Roseville, Swan

Creek, Sciota, Table Grove, Tennessee and Vermont. To me, the names read like the casualty list from a long forgotten war.

Each of these towns has its points of interests and unique characteristics. Industry has a large windmill on the north end of the community. Bushnell still has the Vaughan hammer factory. Colchester holds an enormous Labor Day yard sale event every year that attracts people from throughout the region. But drive through any one of these towns and you will see similar patterns of development and decay. Most of the schools stand empty and abandoned, as do many of the buildings that once housed businesses on the towns' main streets. The largest, most beautiful homes are those that were built many decades ago and many of these are in disrepair. The newer homes, if there are any, are often pre-fabricated or trailers. In several of these communities, the local gas station/convenience store is one of the few successful businesses. Often, it is the only source of food.

It wasn't so long ago that this county was a very different place. My grandparents were born here in the 1910s and have lived in McDonough county all of their lives. They remember a time when the countryside was made up of small fields and farm animals and the people that raised and cared for them. For every mass-produced steel outbuilding that you see on the landscape today there were once many timber-framed wooden barns, each handcrafted and unique: art on the landscape. For every modern day agribusiness person who has agreed to host some outside corporation's feeder pigs, dozens of independent farmers once managed their own outdoor hog lots. For every section of land now owned by outside investors and rented out to a hired equipment operator, several farming families once owned and cared for their own special places; the diverse small farms that they called home.

When my grandfather started farming on his own in the 1930's, many farms were still almost entirely powered by the sun. This fact seems remarkable today, given the fossil fuel dependence of modern agricultural operations and the industrial food system as a whole. In those days, many farmers still plowed, cultivated and/or harvested with teams of horses or mules, the traditional "engines" of the small farm (Figure 1). The horses were fed on oats, grass, hay and other crops raised on site, as were the cows, the pigs, the chickens, and the other animal members of the farm. The manure from these animals was carefully collected and returned to the soil to improve the fertility for future crops; a timeless and elegant arrangement practiced by most of our ancestors right back to the time when agriculture began.

When my grandparents were young, the small towns and communities of McDonough County were lively, vibrant places full of local people and local commerce. On Saturdays, neighboring farmers and their families came to town to get their haircut, to buy supplies and to socialize. On Sundays, the churches were full of local people who gathered to hear the word of God and to meet up with their friends and neighbors.

In my grandparents' day, the most important thing raised on the farm (besides young farmers) was food. This was good food. Quality food. Organic food. It was food raised by hand and lovingly and skillfully prepared in the kitchen. It was food that was produced with pride and care to feed family, friends and neighbors, with the surplus sent out of the community to larger markets. In my grandparents' day, food was raised and prepared using methods that people had learned from their parents; using skills handed down from generations of grandparents and great-grandparents. These were the food and farm skills that had served our ancestors in good times and bad, through periods of wealth and famine from long before our (mostly) European ancestors came here to settle some of the finest agricultural land on earth.

Figure 1: Traditional agriculture with relatives of John Curtis and a hired hand (1915)

My grandparents are both in their mid-90s and their generation will soon move on forever, taking with them their cherished memories of the rich, vibrant communities they knew when they were young. The unsettled question is whether those memories can inspire the change necessary to reverse the degradation of our small towns and rural landscapes and the decline of the people who have stayed behind and still inhabit these places.

It will be argued that I am being naïve here; that I am simply romanticizing a distant past. Agribusiness people and the interests they represent will point out that the realities of a global market have led to the agriculture that characterizes our region today; that industrial scale production is the logical response to the economic realities of our food and agriculture system. They will argue further that agribusiness operations need to continue to reach for ever greater efficiencies of scale and to work even more closely with global corporate interests in order to increase production and expand exports.

These are logical points, especially when you consider the source. Agribusiness people have been forced to think in terms of profits and losses. Any other concerns are secondary. Those who share this mindset have survived and (in some cases) thrived. Those who haven't have been pruned away over time. To a large degree, this process has led to the deserted rural landscapes that surround us today.

However, I think it is becoming clear that the time has come to question the status quo. People are beginning to recognize that those who view modern, conventional agriculture as wholly good or beneficial must be doing so from a narrow perspective; certainly one that ignores the realities on the ground in our part of the world. In other words, we need to look beyond what is profitable for a few as we consider how we move forward into a future that we all share.

If the future of food and agriculture is going to include health as well as profit, then it seems clear that we must begin to focus on raising local food for local people. If our deep, rich, prairie soils are going to be managed to benefit the people who actually live on them, then our agriculture must decentralize and be practiced on many different scales. If there is going to be a resurgence of our small towns and rural communities, then food and agriculture must foster and support local economic systems. If people are going to return to and inhabit our deserted small farms and homesteads, then they must be attracted to these places by bonds of the heart as well as economy.

People sometimes jokingly refer to our part of the world as "flyover country." We live in a region that is considered drab and uninteresting. We don't have the mountains, the beaches or the large bodies of water that might attract people to other parts of the country. You could add that the weather isn't always the most comfortable here. Our summers are often hot and humid and the winters are invariably long and cold.

But, if this is flyover country, then who do we have to blame but ourselves? The original native prairie was a unique and special environment that must have been a remarkable sight. Equally beautiful, to my way of thinking, would have been the great patchwork quilt of small fields, hedgerows and farmsteads of my grandparents' youth. To view a sunrise over that landscape would certainly mean more to me than any visit to some distant tourist destination.

The great tall grass prairie is gone forever. And, as the last vestiges of the agrarian communities that replaced it fade as well, we are faced with a choice. We can choose, by default, to continue moving towards a landscape dominated by machines and chemicals, agribusiness genetics, global capital and outside control of our land and resources. We can choose, by doing nothing, to watch our people become ever more sickly -crippled even- as they consume the tainted, industrial products that get their start right here, on our fields and farms.

But really, there can be no choice. We must act. We must take up the work that will create, nurture and protect our beautiful and healthful landscapes. Or, at the very least, we must support those who will do this work.

In my view, the future of food and agriculture is diverse. It is decentralized, democratic and grassroots. It improves the health of the community, is available to people of all incomes and is ecologically sound. It is time for us to reoccupy and rejuvenate our rural landscapes. It is time for local food.

Chapter 6: Prairie Fruits Farm and Creamery, Champaign, Illinois

Wesley Jarrell and Leslie Cooperband

Introduction

Prairie Fruits Farm and Creamery was established in 2003 as a small fruit farm, to be managed sustainably and organically, part-time, primarily by the owners. In 2013, PFFC has five full-time and seven part-time employees in addition to two full-time owners, milks 70 goats per year, produces nearly 20,000 lbs. per year of artisan goat and sheep cheese, and conducts local food/slow food meals more than 20 times per year. Our growth reflects in some ways the growth of the growing dynamic, energetic, and hopeful local foods segment of our food system. Principles of sustainability continue to guide our growth and decision-making, although we realize we have much to learn and implement. Perhaps our experience can serve to guide other new farmers in developing their own business and rural lifestyle.

Personal Search for Sustainability

In the late 1970's one of my MS students at the University of California-Riverside, Connie Rush, convinced me to attend the Small Farm Conference in La Honda, CA. The conference was held at a Boy Scout camp in the redwoods on the San Francisco peninsula, in February, when farming is even slow in California.

As a University of California Professor, I learned quickly that the organic and biodynamic farmers were not treated well by the University of California, felt like minority clients who were deemed too small and marginal to be worthy of attention. Around that time Jim Hightower wrote a book "Hard Tomatoes Hard Times", discussing among other things the mechanical tomato picker that required a nearly indestructible tomato, which was bred for it. The whole discussion characterizes the state of affairs in California agriculture at that time.

In early 1980's Senator Petris from San Francisco sponsored a successful bill in the California legislature that focused on organic farming initially, changed to energy conservation for political reasons (time of "First Energy Crisis"), and prodded the University of California to create a sustainable agriculture task force. Each of the land grant campuses of UC, which included Berkeley, Davis, and Riverside, were represented, as well as Santa Cruz, with their Agroecology program. In fact, Steve Gliessman at UCSC played a central role in developing the Petris bill, and was a critically important task force member, representing agroecology from a non-land-grant campus, UC Santa Cruz.

Sustainable was code for "organic" in those days, more palatable to legislators and conventional farmers (if just barely). We all thought "small farm" and "organic farm" were nearly synonymous, given how hard it was to be organic.

Organic/Sustainable Agriculture

Although we were dedicating a great deal of energy to working with organic farmers to make their systems more viable, we realized that organic and sustainable were not in fact synonymous. The first person to point that out to me was a California organic farmer who was understood that he was organic, but not sustainable. His farm relied on chicken manure obtained from a large conventional layer farm; he was sure his system could not work without that input. Since then I've encountered numerous cases where organic practices were not truly sustainable and would have to be modified, such as ineffective tillage for weed control and phosphorus build-up from long-term applications of P-rich manure.

Organic and sustainable agriculture are concepts we now have thanks to early pioneers who led the discussion both intellectually and by example. After WWI there was increasing awareness in agriculture that industrial approaches, rather than biological approaches, were beginning to dominate farming. Horses and mules were being replaced by tractors; nitrogen was coming from synthetic chemicals rather than legumes in the rotation or manure from the barn. Without the need to maintain horses for draft, and the option of buying bagged concentrated fertilizer rather than accumulating and spreading manure or planting cover crops, the absolute requirement to have animals on the farm disappeared.

Without the requirement for draft animals, the need for pasture, hay, and animal grains like oats also vanished. The option of grain-only farms became feasible. Before this conversion, skills with biology – feeding and husbanding horses and other livestock, and managing a diversity of crops on the landscape – were essential. As tractors and chemicals came to dominate the farmscape, mechanical and quantitative skills became most important. Farmers who were not able to make this conversion were forced out of agriculture, and holistic agro ecosystem biology basically disappeared.

These changes prompted response from many individuals who were early to identify what was lost in this transition, including the quality of the food and the way people interact with the source of their food. Through their own experience and experiments they came to similar conclusions. The practice of agriculture was reduced into many isolated pieces, rather than an integrated system, with dire consequences for human health, environment, and society.

Sir Albert Howard wrote *An Agricultural Testament*, describing his methods of production in India that recycled waste, provided work, and fed the population. His work inspired many others, among them Lady Eve Balfour, J.I. Rodale, and a host of lesser known agriculturalists.

Lady Eve Balfour founded the Soil Association in England in 1942 and wrote *The Living Soil* (1948). When she addressed the International Federation of Organic Agricultural Movements, 1977, she stated the following: "There are two motivations behind an ecological approach—

- One is based on self interest, however enlightened, i.e. when consideration for other species is taught solely because on that depends the survival of our own.
- The other motivation springs from a sense that the biota is a whole, of which we are a part, and that the other species which compose it and helped to create it, are entitled to existence in their own right. This is the wholeness approach and it is my hope and belief that this is what we, as a federation, stand for" (to IFOAM in 1977).

Pay Dirt, published in 1945, by J.I. Rodale, was written after Rodale had founded *Organic Farming and Gardening* magazine (whose title, after protests from agribusiness, was changed to *Organic Gardening*). In it he describes what he considers the negative aspects of industrial farming systems.

In a 1972 edition of *The Economist*, Blueprint for Survival, the following criteria were laid out for effective new sustainable systems:

- Minimum disruption of ecological processes;
- Maximum conservation of materials and energy;
- A social system in which the individual can enjoy, rather than feel constricted by, these conditions;
- Systematic substitution of the most dangerous aspects of present technology with ones that cause minimum disturbance to natural processes;
- Decentralization of economy at all levels;
- Formation of communities small enough to be reasonably self-regulating and self-supporting

Jerry Goldstein (1976) makes the following statement, which rings as true today as when he first wrote it: "The organic concept is a forerunner of how an economic base can be given to idealistic concepts." Clearly the organic/sustainable concept is an ideal, but we acknowledge at the same time that it must be economically sound to survive without subsidy.

Our farm's story has elements of all these viewpoints. Here it is.

Origins of Prairie Fruits Farm and Creamery

Wes Jarrell was offered the position of Department Head at the University of Illinois – Urbana/Champaign in Fall of 2002. We were academics at the University of Wisconsin-Madison at the time. A friend on the faculty (wife of an organic farmer) at UIUC found the property they eventually purchased. Initially offered at 2.5 acres, with a house and a 5,000 square foot metal-sheathed pole barn, we negotiated the purchase of another 4.5 acres of farmland, for a total of 7 acres (Figure 1).

2003 Purchase 7 acres 2 miles north of Urbana, IL; plant buckwheat cover crop

2004 Plant 350 fruit trees, 200 berry plants. Buy three does and one buck Nubian goat

2005 Licensed as first farmstead cheese plant in Illinois, 25 milkers. First sales at Urbana Farmers' Market

2006 50 milkers

2007 Grant from Frontera Farmer Foundation, build commercial kitchen; access to 15 acres east edge of farm (Illinois Farm to Fork) plant pasture/hay/prairie

2008 First five Dinners on the Farm; lease 17 acres for hay

2009 First Breakfasts on the Farm; 60 milkers; 12 dinners, Sheep milk from Plank fair Arthur, IL

2010 Lease 20 additional acres for hay; 70 milkers; 12 dinners

2011 Learn gelato-making in Bologna, sales on farm and at Urbana Market; 70 milkers; 14 dinners

2012 Cheese and gelato CSA; First prize at American Cheese Society; 70 dinners; 14 dinners, plus special events

Figure 1: Timeline for Prairie Fruits Farm and Creamery

The Fruit Farm

The original vision was a fruit farm; we wanted to pick fresh peaches off the tree when they were dead ripe and eat them in the field or can them. And they wanted to share the experience with others who may have never had perfect fruit. So in spring 2004, they planted 350 fruit trees (10-12 trees each of about 30 varieties) including peaches, apples, pears, cherries, as well as over 200 berry and currant plants. They started with organic management and have maintained it for the past 10 years.

The Milking/Goat Test

In summer 2004, we purchased three Nubian goat does and one buck. Before they separated the buck and does, two does became pregnant, unbeknownst to the owners. These two sisters, Snickerdoodle and Chocolate, gave birth right after Thanksgiving, three months earlier than we'd planned. We milked them in the dirt-floored pole barn all winter, and made cheese from the milk in our house kitchen. All our friends and relatives got to taste our many efforts with the cheese. Pronouncing it good, we made the plunge and went ahead with development of a grade A dairy and manufactured milk plant (cheese plant).

Goat Dairy

After consulting with the Illinois Department of Public Health, we began construction of the milking parlor and the milk house in the pole barn (5000 square feet with a dirt floor). We found we needed licensed plumbers and electricians to install everything. Based on the projected volume of water waste generated, we build a septic system of three 1500 gallon tanks and drain field. To milk initially we used a "bucket" milker, a bucket with inflations (two, one for each teat on the goat) connected to a vacuum pump. We then filtered the milk through a paper filter into a refrigerated bulk tank. Our two bulk tanks (80 and 100 gallon) had been recovered from the back of two barns in Wisconsin.

With 25 milking goats milked twice a day, this took quite a bit of time. When we moved about 35, we worked with a dairy company to install an enclosed stainless steel "pipeline" system that lets us milk and clean the system much more efficiently.

Cheeserie

We purchased a 200 L (52 gallon) vat-pasteurizer from a company in the Netherlands. Two years later we purchased another. Interestingly, after looking at used systems, we realized the cost was nearly the same for high quality used equipment as it was for new equipment, even counting the import costs.

We initially made only pasteurized cheeses, fresh chevre and bloomy rind cheeses (Camembert and Brie styles) we could age in containers in our walk-in cooler for two to four weeks before selling. After another year, we built four small aging rooms where we could, more or less, control temperature and humidity to produce aged raw milk cheeses as well as our aging bloomy rind cheeses. Raw milk cheeses must be aged more than 60 days before sale.

Initially all our sales went to the Urbana Farmers' Market, which we have since discovered is one of the best markets in the state of Illinois. Gradually we built up markets with direct wholesale to restaurants and grocery shops and stores, and eventually distributors (Figure 2). At the current time our sales are split between direct (farmers' markets, CSA, and on-farm), wholesale, and distributors. Of course we'd prefer to sell all at the highest markup, but this is not practical because of time and logistics concerns.

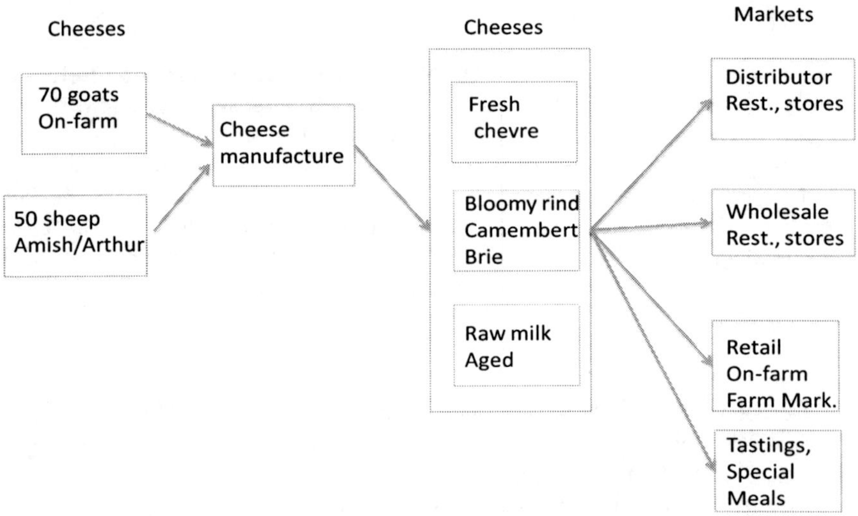

Figure 2: Flow of milk from Prairie Fruits Farm and Creamery

Add Sheep Milk

In 2006 we met two Amish brothers-in-law who wanted to raise and milk sheep. They asked if we'd buy their milk. We finally agreed that, if they could meet our quality standards, we would pay them a good price for the milk.

In 2007 they brought 25 ewe lambs, each from a farm in Wisconsin. After several fits and starts, one of them was successful in starting a sheep dairy (the first licensed sheep dairy in Illinois). We travel to their farm, about 50 minutes' drive south of us near Arthur, Illinois, twice a week, picking up 50 gallons each time.

Sheep milk is 50% richer than our goat milk, both butterfat and protein, so makes significantly more cheese per gallon than goat milk. It has some other characteristics that require special attention at times, including the propensity to grow a type of black mold at certain times of the year. Sheep generally milk only about 6 – 7 months per year, starting about the same time as the goats.

Dinners on the Farm

Because we had organic fruit, which tastes extremely good but isn't always perfect-looking, we'd always thought that we would develop a system for jams, jellies, and preserves of various types using our fruit. Leslie and our Associate Cheesemaker (now our chef) Alisa Demarco wrote a proposal to the Frontera

Farmer Foundation (http://www.rickbayless.com/foundation/about.html) for about 1/3 ($12,000) of the cost for building and outfitting a commercial kitchen and we came up with the rest. In summer 2008 our kitchen was certified by Champaign County, meaning that we could use it to prepare foods for sale.

We had hosted several dinners on the farm as potlucks, the diners from local churches to small groups of horticulturalists. The most memorable was a farmer potluck with Carlo Petrini. Mr. Petrini, founder of Slow Foods International, was on a US book tour and wanted to meet Illinois farmers, so in May 2008 we brought together about a half dozen other farmers and hosted him for a dinner on our farm. He insisted that everyone sit at a single long table. It was a transformative experience and informed our decision to begin offering slow food/local food dinners on the farm.

After our kitchen was licensed, we ran a trial meal with friends. Other than a mosquito attack and the ineptitude of the servers (the farmers Leslie and Wes), it went very well. Alisa, a trained chef with a degree from the Culinary Institute of America in New York and experience at restaurants in Austin, Chicago, and Champaign, has a highly developed sense of ingredients as well as all professional chef skills- she knows how to turn out superior products.

In 2008 we hosted five dinners on the farm with guests. Forty-five to fifty guests are seated at two long tables. Reservations are made on the web. We alert our email list about when the dinners will go up on the website. About 14 dinners a year are offered, starting in late May and running through early December. Reservations were made for forty, with the two farmers sitting down with the guests. Typical hours are 4 – 8 pm on Saturday evening. Guests are sent the menu a few days ahead of time and typically select wines or beers to complement the courses.

When guests arrive, they are greeted with three different hors d'oevres over bout thirty minutes. Then the farm tour begins and continues for 30 – 35 minutes, discussing and visiting the garden, fields, and goats. Guests are encouraged to ask questions and offer suggestions. At the end of the tour, guests return to the dining area (a slightly raised wooden platform outside in good weather, a finished dining area in the barn in bad weather) and select their seats.

Chef Alisa then does a presentation, describing each course, the source of the ingredients, and any special techniques she's used in the preparation of the dishes. Guests can ask questions at any time. We introduce the guest farmer at that time, who presents a short description of the farm and the featured products from the farm. The farmer is available throughout the meal to answer questions from the guests.

The main dinner event is usually four to five courses, including a cheese course and dessert. Guests are free to walk around the farm between courses, taste vegetables or fruit, and ask questions of the farmer-owners. Chef Alisa is always available in the kitchen to answer any questions.

Just before 8 pm, all of the kitchen staff assembles in the dining room, are introduced by Chef Alisa, and thanked by the guests. Guests typically disperse over the next half hour.

Dinners are only advertised on our website. Typically four or five dinners are sold at a time, and sell out in 10 minutes or so.

We feel the following are the basis for their popularity:

- Chef Alisa is an excellent professional chef and has a deep and thorough understanding ingredients, agriculture, gardening, staff, and guests.
- Farmers/owners/hosts make sure guests are comfortable, their questions are answered, and that they get a true farm experience in addition to the excellent meal.
- Goats are charismatic animals, being similar in size and sometimes behavior to humans. Ours are raised to be very comfortable around people, inviting their attention and occasionally providing interesting moments (like opening the gate and escaping, milling about during the meal!).

Breakfasts on the farm

Starting in 2006, we had a started a small, informal farmer's market at the farm, inviting a couple of other farmers and a baker who had early season products, before the regular farmers market started in May. Once we had the kitchen licensed, and all these excellent products, we decided to offer a breakfast option for visitors in March and April. Very rapidly the experience morphed into a breakfast experience at which we sold some cheese and vegetables. We now regularly see 300 guests between 9 and 12 on Saturday morning. Most buy something, although a few just come to see the baby goats – which is fine with us.

Gelato

In line with our desire to make best use of synergies on our farm, we were faced with the question of what to do with our surplus milk in spring, right after kidding. Gelato seemed a good option, so we purchased a used machine, experimented with recipes and production, then went to Bologna, Italy, for a one week course on gelato preparation. We were focused on using as many local ingredients as possible, so we felt we needed to understand the basics of gelato preparation, production, and serving.

In spring 2011 we started regular production of gelato, selling on the farm and at farmers' market in Urbana. Using as much of our own fruit and herbs as possible, we created popular flavors in spring and through the summer.

Tours/Education

Since we both came from a strong background in education, specifically soil science and agriculture, it's been natural for us to include more and more of those opportunities in our business. The tour portion of the on-farm dinners includes an overview of our farm, its history, visits with the goats and walks through the garden when the weather is agreeable. We make sure the guests know that we're happy to answer any questions they may have about our operation.

Recently we have begun formalizing our tour offerings. We have been charging $5 per student for tours up through college, with a $100 minimum. For adults we charge $8 per person with the same minimum. For tours of more than 30, we break into two groups.

For each tour different portions of the farm will be emphasized, depending on the interests of the tour guests. All tours end with a tasting of cheese, gelato, or both, again depending on the interests of the attendees.

Conclusion

Sustainability is much more about the journey than the destination, as many have said. That's because the external environment – economic, ecological, and social – changes continuously, meaning that any reasonable approach to survival has to take these changes into account on a regular basis. The approach that may have worked ten years ago will probably need to be adjusted, sometimes radically, to work today.

The regular holistic analysis of the external world and the way the operation works within it distinguishes the sustainable operation from the conventional. These are not new ideas.

Sir Albert Howard, writing in the Introduction to *Pay Dirt* (1945), says: "Everywhere knowledge increases at the expense of understanding. The remedy is to look at the whole field covered by crop production, animal husbandry, food, nutrition, and health as one related subject and then to realize the Great Principle...the birthright of every crop, every animal, and every human being is health."

We feel that Sir Albert got it right with his analysis and we want to work with other like-minded individuals – farmers to eaters – to help society achieve this birthright, melding idealism embodied in the pronouncement with the practicality of economic forces.

References

Balfour, Lady Eve. 1943. *The Living Soil*. New York: Universe Books.
Howard, Sir Albert. 1943. *An Agricultural Testament*. New York: Rodale Press.
Merrill, Richard. 1976. *Radical Agriculture*. New York: HarperCollins.
Rodale, JI. 1945. *Pay Dirt*. New York: Rodale Press
Rodale, Maria. 2010. *Organic Manifesto*. New York: Rodale Press.

Chapter 7: Legal and Institutional Issues in Local Food Production: Lessons from Illinois

A. Bryan Endres, Lisa Schlessinger, Rachel Armstrong

Introduction

The burgeoning reorientation of the food supply toward more locally sourced products illustrates the frustration of many consumers regarding our industrialized agricultural system—a form of agriculture that has replaced personal connections between farmers and their community with large scale, highly structured production and logistical operations dependent on economies of scale, high volumes of off-farm inputs and complex transportation networks—developed to provide an the population with highly-processed food.[1] The desire for a more connected experience to one's food and a lowered impact on the environment is a common motivation underlying consumer demand for a more local food system.[2] While federal farm bill and state agricultural promotion definitions of what may qualify as "local" often focus simply on state borders or the distance a food travels between the farm and retail outlet, a more nuanced food system assessment should take a broader account of the production, distribution and consumption of food, from the time it is planted in the field until it reaches the consumer's fork.[3] From this perspective, a "local" food product embodies more than distance, to include some form of connection to a community, broadly defined.[4] In addition to purchasing decisions, communities

---

[1] Nicholas R. Johnson & A. Bryan Endres, "Small Producers, Big Hurdles: Barriers Facing Producers of 'Local Foods'," *Hamline Journal of Public Law and Policy* 33 (2011): 50-51; A. Bryan Endres & Jody M. Endres, "Homeland Security Planning: What Victory Gardens and Fidel Castro Can Teach Us in Preparing for Food Crises in the U.S.," *Food & Drug Law Journal* 64 (2009): 428-429; Mark Bittman, "Fixing our Food Problems," *NY Times* (Jan 1, 2013), available at http://opinionator.blogs.nytimes.com/2013/01/01/fixing-our-food-problem/.

[2] Derrick Braaten & Marne Coit, "Legal Issues in Local Food Systems," *Drake Journal of Agricultural Law* 15 (2010): 10.

[3] Ibid., 13-14.

[4] Megan Galey & A. Bryan Endres, "Locating the Boundaries of Sustainable Agriculture," *Nexus Chapman's Journal of Law & Policy* 17 (2012): 4-5.

may express shared values through their legal systems.[5] Statues and regulations embody what the community collectively chooses to prohibit, protect or subsidize. Accordingly, it is critical to examine the intuitional biases within the current agricultural production system established through the legislative and regulatory process, and its residual impact on the development of local food systems.

An enormous web of tangled federal and state laws attempt to simultaneously regulate and exempt many local food producers from the larger landscape of food safety and operational regulations that oversees food production in the United States. These complex and conflicting rules unnecessarily impose barriers to the expansion of local food systems.[6] In addition, institutional issues—such as financing and land procurement, as well as labor rules designed for industrial production rather than small scale, direct market farming—pose threats to the viability of farmers seeking to shorten food supply chains and directly market products to consumers.[7] Moreover, these novel issues also present challenges to rural lawyers with expertise in representing commodity agricultural producers, as well as urban attorneys unfamiliar with the intricacies of food production and who might find it frustrating to handle a farmer's needs during a nine-to-five schedule.[8] Accordingly, this chapter will introduce the array of the legal issues confronted by farmers engaged in direct marketing, and identify farmer-focused solutions to overcome the barriers to expanding opportunities for localized food sheds. The analytical framework focuses on the farmer seeking to directly market products to consumers. As a foundation of an expanded local food system (e.g., farmers' markets, roadside stands, u-picks, community supported agriculture programs, as well as direct sales to restaurants, grocery stores, and schools) these farm-entrepreneurs embody the personal connection between the growing field and food increasingly demanded by consumers.

Scale and Complexity

A hierarchy of laws, including federal, state, county and even municipal rules governs food production and labeling.[9] Federal laws preside over any food that enters channels of interstate commerce, meaning travels from one state to another.[10] Although at first glance this may not appear to have a significant impact on a farmer trying to sell directly to a local resident, courts have interpreted the federal government's authority under the Commerce Clause to include activities that take place entirely within a state (intrastate), if the aggregate of those intrastate activities have substantial effects on commerce involving multiple states.[11] This would include diverting to purely intrastate

[5] See Generally, H.L.A. Hart, *The Concept of Law* (Oxford University Press, 1961).
[6] Nina W. Tarr, "Food Entrepreneurs and Food Safety Regulation," *Journal of Food Law and Policy* 7 (2011): 36.
[7] Kenneth Miller & Adam Prizio, "Innovate Use of the Law for Small-Scale Producers," *Journal of Environmental Law and Litigation* 26 (2011):136-37.
[8] Ibid., 133.
[9] Tarr, *Food Entrepreneurs*, 36.
[10] United States v. Darby, 312 U.S. 100 (1941).
[11] Wickard v. Filburn, 317 U.S. 111, 124 (1942).

commerce food products that otherwise might flow in interstate commerce or replace the purchase of goods flowing in interstate commerce.[12] In light of the current national (and international) market for virtually every food product, Congress has authority to regulate all aspects of food production, distribution and marketing.[13] The result is fifteen separate sub-agencies of the federal government responsible for implementing approximately thirty federal statutes relating to food safety along with implementing regulations for each statute.[14] Most of these statutes seek to regulate (or in some cases subsidize) the commodity-based, industrialized agriculture system advocated by a series of USDA officials, succinctly summed-up by Secretary of Agriculture Earl Butz's infamous admonition to "Get big or get out!"[15]

For the small scale food entrepreneur the sheer number of federal laws that may apply is overwhelming.[16] Moreover, in response to bioterrorism threats and several high-profile food borne illness outbreaks, many new regulations have increased oversight of food production.[17] For example, in accordance with the Public Health Security and Bioterrorism Preparedness and Response Act of 2002 (the Bioterrorism Act), the FDA added a rule to the Federal Food, Drug, and Cosmetic Act (FFDCA) requiring all facilities that hold, pack, manufacture or produce food for animal or human consumption in the U.S. to register with the FDA prior to beginning manufacturing/processing, packing or holding food.[18] The purpose of facility registration is to help the FDA to determine the location and source of potential bioterrorism incidents or outbreaks of food-borne illnesses.[19] Generally, farms are exempted from registration requirements,[20] however if a farm does more than grow and harvest crops, or raise animals (*e.g.*, processing food for consumption off the farm), then the farm must register.[21] Additionally, if a farm holds foods from other farms with different ownership, then the farm must register.[22]

These outbreaks are not just a processed food problem. For example, in 2006, the FDA traced an *E. coli* outbreak to spinach; representing the first time the agency established a link between farm fields and bacterial contamination.[23]

---

[12] Gonzales v. Raich, 546 U.S. 1, 17-22 (2005).

[13] Johnson & Endres, *Small Producers*,120.

[14] Michael T. Roberts, "Role of Regulation in Minimizing Terrorist Threats Against the Food Supply: Information, Incentives, and Penalties," *Minnesota Journal of Law, Science & Technology* 8 (2007): 205.

[15] A. Bryan Endres & Nicholas R. Johnson, "Integrating Stakeholder Roles in Food Production, Marketing, and Safety Systems: An Evolving Multi-Jurisdictional Approach," *Journal of Environmental Law & Litigation* 26 (2011): 40.

[16] Johnson & Endres, *Small Produers*, 121.

[17] Ibid., 123.

[18] 21 U.S.C. § 350(d) (2012); 21 C.F.R. pt. 1.227.

[19] 21 U.S.C. § 350(d) (2012).

[20] 21 C.F.R. § 1.226(b)(3) (2012) ("Farm means a facility in one general physical location devoted to the growing and harvesting of crops, the raising of animals, or both. Washing, trimming of outer leaves of, and cooling produce are considered part of harvesting").

[21] 21 C.F.R. § 1.225(2012); 21 C.F.R. § 1.226(b)(3)(ii)(2012).

[22] 21 C.F.R. § 1.225 (2012); 21 C.F.R. § 1.226(b)(3)(i) (2012).

[23] Johnson & Endres, *Small Produers*,124.

Similarly, *Salmonella Enteritidis* (SE), a contaminant that affects unprocessed foods, has been associated with public health implications since 1988.[24] In response to these outbreaks, the government has stepped up its regulation of unprocessed foods. While the USDA oversees egg quality, the FDA is responsible for the safety of shell eggs throughout the supply chain. In 2009, the FDA implemented an SE eradication program.[25] The law requires all but the smallest shell egg producers, those with over 3000 laying hens, to implement measures to prevent SE from contaminating eggs during production, storage, and transportation.[26] The measures include writing a SE prevention plan, limiting visitors for bio-security purposes, and controlling for rodents and pests.[27] Producers must also perform environmental SE testing, and if found, the producer must expand the testing to the actual eggs.[28]

In the wake of the 2006 *E. coli* outbreaks, both industry and the government set their sights on the safety of leafy green produce. The U.S. industrial-scale food system produces unique challenges to safety monitoring because food companies commonly consolidate and commingle produce from farms across the U.S. at a few regional processing plants, which bag the greens for extended shelf life, and then ship them across the country.[29] This system makes trace-back after an outbreak extremely challenging and places a large number of consumers at risk from a single contamination event.[30] Aside from expected regulations arising from passage of the Food Safety Modernization Act, the FDA has refrained from issuing any binding regulations dealing with fresh produce safety. Rather, the leafy green produce industry proposed a voluntary public-private marketing agreement among industry players and regulators.[31] Marketing agreements are government sanctioned requirements that bind growers, handlers and processors to establish best practice metrics for the safe handling of a product.[32] Although technically voluntary, compliance with an agreement may be a necessary precursor to market access and, once a party signs the agreement, the terms are binding and mandatory.[33] In 2009, a group of leafy green trade associations petitioned the USDA, asking for a national leafy green marketing agreement (NLGMA) applicable to all leafy green handlers in the U.S., under the authority of the Agricultural Agreement Act of 1937.[34]

---

[24] U.S. GEN. ACCOUNTING OFFICE, GAO/RCED-99-184, FOOD SAFETY: U.S. LACKS A CONSISTENT FARM-TO-TABLE APPROACH TO EGG SAFETY 2 (July 1, 1999), *available at* http://archive.gao.gov/f0902b/162324.pdf.

[25] 21 C.F.R. § 118 (2012).

[26] 21 C.F.R. § 118 (2012).

[27] 21 C.F.R. § 118.4 (2012).

[28] 21 C.F.R. §§ 118.5 and 118.6 (2012).

[29] A. Bryan Endres & Nicholas R. Johnson, "Integrating Stakeholder Roles in Food Production, Marketing, and Safety Systems: An Evolving Multi-Jurisdictional Approach," *Journal of Environmental Law and Litigation* 26 (2011): 58.

[30] Ibid.

[31] Ibid. At 66.

[32] Ibid.

[33] Ibid.

[34] Proponent Group's Brief in Support of a Federal Marketing Agreement for Leafy Green Vegetables at 2, Leafy Green Vegetables in the United States: Hearing on Proposed

One may question why industry itself would push for stricter, and costly regulations of their own product, as the leafy green trade association did. In addition to obvious public health concerns and potential liability for the farm/processor at the source of the contamination, food borne illness outbreaks impose significant costs to the broader industry as consumer panic and precautionary public health warning may implicate an entire commodity—shutting down an industry and suppressing demand for extended periods after the outbreak. In response, large industry interests have sought ever more complex technological solutions codified into regulations. Unfortunately, compliance with these marketing agreements or regulations may prove financially out of reach to the small scale direct farm business.[35]

In 2011, the federal government enacted the Food Safety and Modernization Act (FSMA)[36], which may prove to be the most extensive overhaul of the nation's food safety regulatory authority since the FFDCA in 1938.[37] The FFDCA is the primary food safety statute at the federal level[38] and prohibits the introduction into interstate commerce of adulterated or misbranded food and authorizing criminal and civil penalties.[39] The FSMA amendments to the FFDCA should modernize food system regulation through increased oversight; the precise and potentially harmful effect on direct farm operations remains to be seen, however, until after the FDA promulgates the implementing regulations.

Broadly, once fully implemented, the FSMA will shift regulatory focus away from *ex-post* crisis management towards *ex-ante* prevention of foodborne outbreaks.[40] The FSMA intends to achieve this objective by requiring Hazard Analysis and Risk-based Preventative Controls (HACCP) for foods under FDA jurisdiction.[41] HACCP requires design and implementation of procedures to assess, monitor, and correct aspects by which food may become contaminated. The FSMA also authorizes the FDA to impose production practices at the farm level.[42] These standards will detail specific production and harvesting standards to minimize food safety breaches. The FDA also may prioritize certain individual standards for high-risk produce[43]—a step that the federal government up to this point has been reluctant to take. High-risk facilities, which may include farm processors, are also subject to FDA inspection under FSMA.[44]

Marketing Agreement No. 970 Before the U.S. Dep't. of Agric., No. AO-FV-09-1038, *available at*
http://www.ams.usda.gove/AMSv1.0/getfile?dDocName=STELPRDC5082273.

[35] Ibid.,123; *see* A. Bryan Endres, "United States Food Law Update," *Journal of Food Law and Policy* 3 (2007): 104-07.

[36] P.L. 111-353, Jan. 4, 2011, 124 Stat. 3885.

[37] Ibid., 128.

[38] *See generally* Federal Food, Drug, and Cosmetic Act, 21 U.S.C. § 331 (2006).

[39] *See* 21 U.S.C. § 342(a) (2006) (defining adulterated food); 21 U.S.C. § 343(a) (2006) (defining misbranded food).

[40] Tarr, *Food Entrepreneurs*, 41.

[41] Section 103, P.L. 111-353, Jan. 4, 2011, 124 Stat. 3885.

[42] Section 104, P.L. 111-353, Jan. 4, 2011, 124 Stat. 3885.

[43] Section 105, P.L. 111-353, Jan. 4, 2011, 124 Stat. 3885.

[44] Section 201, P.L. 111-353, Jan. 4, 2011, 124 Stat. 3885.

Production facilities that pose a reasonably probable risk to public health may have their registration suspended, a cementing of authority the FDA had previously not held over farm production.[45] Additionally, the FSA authorizes mandatory recall authority to replace the nation's previously voluntary recall system.[46] To effectuate a recall, food facilities are required to keep certain records documenting the production and distribution of products.[47]

Because of the potentially high cost of compliance with the array of new rules expected in the near future as the agency implements the statute, the continued development of local foods systems will be dependent upon the continuation of exemptions for small, direct farm businesses. Congress, thus far, has exempted small producers from many federal food regulations, so long as the farming operation meets certain requirements. For example, retailers with gross annual food sales less than $50,000 are exempt from the FDA's nutrition labeling requirements.[48] The FSMA includes some exemptions for small farms with less than $500,000 in total sales, if half of the total farm sales are direct-to-consumer or restaurant sales within the state or a 275 mile radius.[49] As with the maze of federal rules, qualifications for the various exemptions are complex and vary among programs and implementing agencies (e.g., USDA versus FDA).

Although essential for the survival of many small farm entrepreneurs, the various exemptions exemplify the paradox faced by the local food system. If the movement envisions direct farms (e.g., farmers' markets, CSAs, roadside stands) as the predominant characteristic of the new food system, then many more farms, including many more large farms, are needed. Yet, a regulatory threshold of $500,000 in total sales (half of which must be of the direct-to-consumer nature) provides an incentive for some operations to remain relatively small. Accordingly, farmers need clarity with respect to compliance with the regulations that accompany their growth—the absence of which could result in the local food movement stalling out before obtaining its potential.

Many states, including Illinois, view the development of local food systems not simply as a way to encourage healthy eating, but as a key economic driver for rural communities.[50] In 2007, the Illinois legislature passed the Illinois Food, Farms, and Jobs Act after finding that Illinois consumers spend $48 billion annually on food, most of which is grown and produced outside of Illinois.[51] The Act created the Illinois Local and Organic Food and Farm Task Force, which was tasked with developing policy and funding recommendations for expanding and supporting the production of local foods with the intention of creating jobs and promoting economic growth in rural communities.[52] The task force produced a report recommending the creation of the Illinois Local Food, Farms, and Jobs

---

[45] Section 102, P.L. 111-353, Jan. 4, 2011, 124 Stat. 3885.

[46] Section 206, P.L. 111-353, Jan. 4, 2011, 124 Stat. 3885.

[47] Section 204, P.L. 111-353, Jan. 4, 2011, 124 Stat. 3885.

[48] 21 C.F.R. §§ 101(j)(1)(2011) and 21 C.F.R. § 101.36(h)(1)(2011).

[49] 21 U.S.C. § 419(f) (2012).

[50] 505 ILL. COMP. STAT. 84/5 (2012).

[51] The Illinois Local and Organic Food and Farm Task Force, *Local Food, Farms & Jobs: Growing the Illinois Economy*, 13 (2009), *available at* http://www.agr.state.il/us/newsrels/taskforcereport-outside.pdf.

[52] Megan Galey & A. Bryan Endres, *Locating the Boundaries*, 26.

Council (Council), which would facilitate local farm and food development throughout Illinois.[53] In 2010, the Illinois General Assembly established the Council in order to spur the growth of an Illinois-based local farm and food product economy that would revitalize communities, promote healthy eating, create jobs, ensure a supply of safe food, and support economic growth by making local food products available to all Illinois citizens.[54] The Council is responsible for facilitating the mission of the Act by acting as a liaison between agencies, producers, lawyers and the community.[55] The Act is known as the Local Food, Farms, and Jobs Act, and also contains state procurement goals for Illinois.[56] According to the Act, by 2020, twenty percent of the food purchases made by state agencies, and ten percent of food purchases from entities that receive state funding and spend more than $25,000 per year on food should be local farm and food products.[57]

Yet even in states with active local food initiatives such as in Illinois, overly broad regulatory structures can impede food entrepreneurism. For example, the FDA publishes the Food Code—a model set of rules for state and other governments to enact that provides state and local actors with scientifically sound technical regulations for retail and food service industries.[58] Revised every four years, 48 states and territories have adopted at least parts of the FDA's Food Code,[59] which thus forms the basis for many state laws that govern small producers.[60]

While the Food Code sets out important general food safety parameters for food processing, restaurants and retail establishments, the details of implementing regulations are left to the state agencies—in Illinois, the Department of Public Health (IDPH)—which places inspection and enforcement powers in local public health departments and their inspectors.[61] Illinois has also enacted a state version of the FFDCA that prohibits the sale of misbranded or adulterated food within the state of Illinois.[62] In sum, the Food Code and Illinois Food, Drug, and Cosmetic Act, along with several other state statutes, establish extensive licensing and labeling requirements under a one-size-fits-all regulatory paradigm that can make compliance costly and difficult for the farm entrepreneur.[63]

---

[53] Ibid.; The Illinois Local and Organic Food and Farm Task Force. *Local Food, Farms & Jobs: Growing the Illinois Economy*, 39 (2009). *available at* http://www.agr.state.il/us/newsrels/taskforcereport-outside.pdf.

[54] 30 ILL. COMP. STAT. 595/15 (2012).

[55] 30 ILL. COMP. STAT. 595/20 (2012).

[56] 30 ILL. COMP. STAT. 595/10 (2012).

[57] Ibid.

[58] U.S. Food and Drug Administration, FDA Food Code, http://www.fda.gov/food/foodsafety/retailfoodprotection/foodcode/default.htm (last visited November 12, 2012).

[59] Johnson & Endres, *Small Producers*,126.

[60] Ibid.

[61] Johnson & Endres, *Small Producers*,129. *See* Tarr, *Food Entrepreneurs*, 44 (how Illinois law regulates farmers' markets).

[62] The Illinois Food Drug and Cosmetic Act (410 ILCS 6120).

[63] Tarr, *Food Entrepreneurs*, 36-37. Johnson & Endres, *Small Producers*, 137.

One section of the Food Code, particularly relevant to small producers, provides that all foods sold to consumers, except raw, unprocessed agricultural commodities, must originate from an inspected and licensed facility.[64] As a result, jams, jellies, bread and other seemingly harmless products capable of production by small-scale entrepreneurs need to be made in a licensed commercial kitchen—a significant entry barrier for local food producers.[65]

As at the federal level, some states have crafted exemptions for direct market sales of processed foods, commonly referred to as Cottage Food laws. Cottage Food laws are regulatory carve-outs that exempt certain small food producers from licensing and inspection requirements otherwise provided by the Food Code.[66] Cottage food laws vary by state, but typically share four commonalities: (1) the exemption applies only to producers who operate out of their home; (2) only "non-hazardous" foods (i.e., foods that have very little chance of causing a food borne illness) are exempt; (3) the exemption applies only to small-scale producers as defined by gross sales; and (4) requires specific labeling in order to inform potential consumers that the item offered for sale is not regulated.[67]

Illinois adopted its Cottage Food law in August of 2011.[68] The statute applies to home-based food operations that sell non-potentially hazardous products at farmers' markets. Although the statute incorporates the FDA Food Code's definition of non-potentially hazardous food, it also enumerates specific foods as exempt, including baked goods, jams, jellies and herb and tea blends.[69] To ensure application to only small-scale businesses, the exemption applies to producers with have gross receipts less than $25,000 per calendar year.[70]

Cottage Food laws and other scale-based regulatory exemptions provide important incentives for start-up operations handling non-potentially hazardous foods produced in the home. But concerns that food safety should be regulated even at the smallest producer levels have not gone unheard.[71] Moreover, many successful food entrepreneurs worry that new businesses need an incentive to move out of home-based kitchens into larger commercial facilities in order to thrive in a competitive business environment.[72] Cottage Food exemptions, unfortunately, may curtail that initiative.[73] Nonetheless, Cottage Food laws, and similar federal exemptions, provide a useful tool for a narrowly defined set of producers. It is important to keep in mind, however, that these exemptions, despite their populist media appeal, do not apply to the vast majority of the direct

---

[64] Johnson & Endres, *Small Producers*, 126; U.S. Food and Drug Administration, FDA Food Code, http://www.fda.gov/food/foodsafety/retailfoodprotection/foodcode/default.htm (last visited November 12, 2012).

[65] Johnson & Endres, *Small Producers*, 127.

[66] Johnson & Endres, *Small Producers*, 153.

[67] Tarr, *Food Entrepreneurs*, 58-66.

[68] 410 Ill. Comp. Stat. 625/4(b)(2).

[69] 410 Ill. Comp. Stat. 625/4(b)(1)-(3).

[70] 410 Ill. Comp. Stat. 625/4(b)(5)

[71] Johnson & Endres, *Small Producers*, 156.

[72] Erik Neumann, Homemade Food Act Passes, but Small Producers Aren't Celebrating, *Mission Local* (2012), http://missionlocal.org/2012/10/homemade-food-act-passes-but-small-producers-arent-celebrating/.

[73] Ibid.

farm businesses that are necessary to build and sustain a thriving local foods economy. Small-scale regulatory relief can only go so far. The rest remain subject to the vast umbrella of federal and state legislation.

Finally, it is critical to note that all operations in the business of producing food, regardless of size or qualification for a particular regulatory exemption, are subject to the fundamental prohibition at the state and federal level from selling adulterated food. Thus an otherwise exempt "Cottage Food" operation selling strawberry jam or another product will be held responsible for any foodborne illness. Penalties could include government sanctions, as well as private liability actions brought by injured persons (i.e., victims of a foodborne illness). At the end of the day, there is no regulatory flexibility from the basic principle that it is unlawful to sell adulterated food. Although no producers intend to sell adulterated food, substantial work needs to be done at the farm level to improve regulatory knowledge and craft solutions to enable better navigation of the food safety legal maze burdening the more rapid development of local food systems. [74] Unfortunately, regulatory issues related to food safety present only one of the many legal and institutional issues confronting development of a more vibrant local food economy. In the section below, we explore a variety of other issues from a legal perspective.

Institutional Issues Involving Leases, Financing, and Employment Laws

Access to land remains one of the biggest hurdles facing farmers who want to start a direct farming operation.[75] Purchasing quality farmland is an expensive proposition and the ability to rent land for a value-added direct farming operation presents additional challenges.[76] Absentee farmland owners logically seek to maximize the financial return on their investment through annual leases at the market-rate. Most often, competition for access to rented farmland will be in the form of commodity producers seeking short term (annual) profits with a proven business model that is easy for the landowner to evaluate (e.g., yield per acre less production costs along with any crop subsidies and insurance). The landowner can quickly evaluate the ability to pay a cash rent or assume the risk of a crop-share leasing arrangement. In contrast, the direct farm entrepreneur presents a novel business model, generally will lack crop insurance or other production subsidies,[77] and may require a unique lease arrangement to justify investment in specialty production such as irrigation or cold storage facilities to quick-chill harvested produce.[78] Moreover, many special growers invest heavily in soil health and sustainability practices that also necessitate lease stability.[79] These

[74] Johnson & Endres, *Small Producers*, 131.
[75] Miller & Prizio, *Innovate Use*, 136.
[76] Ibid., 136.
[77] Susan A. Schneider, *Food, Farming, and Sustainability: Readings in Agricultural Law*, 103 (2011).
[78] Equity Trust, Inc., *Preserving Farms for Farmers: A Manual for Those Working to Keep Farms Affordable* (2009): 2, 4.
[79] Edward Cox, *The Landowner's Guide to Sustainable Farm Leasing* (2010): 5-12.

sustainability practices may reduce short-term financial profitability in favor environmental gain; a concept generally not recognized in traditional farm leases.[80]

Unfortunately, due to the complexities outlines above, leasing farmland may not be an option for many beginning farmers seeking to start with direct farm business model.[81] Yet, financing a land purchase presents its own set of challenges. As in the non-farmer landowner situation described above, bank officers administering federal loan programs often are unfamiliar with non-traditional farming operations and vary in their willingness to provide financing to direct farm marketing operations.[82] Some entrepreneurial farmers, as an alternative to banks, have sought community support to raise operating capital by selling "shares" to members in exchange for a future return of goods or services.[83] The most popular form of community supported financing in the agricultural setting is community supported agriculture (CSA), which allows farmers to sell shares of their crops at the beginning of the season. In return, members receive deliveries of produce throughout the growing season based on the number of shares purchased.[84] Potential complicating issues in this business model include how to handle share distribution to farm members in the event of crop failure and, from a more positive perspective, the disposition of excess crops in the event of a highly successfully growing season. These issues, and other potential liabilities, should be worked out in advance with a contractual ("Member") agreement between the farmer and the consumer.[85]

While CSAs may be useful for raising initial capital and/or accelerating cash flow for annual operating expenses, the CSA and other types of direct farming operations—such as famers' markets, roadside stands and u-picks— present unique legal issues and potential liabilities. Many CSAs encourage members to volunteer or intern on the farm. But whether state and federal employment laws would classify these volunteer and interns as "employees" subject to minimum wage, workers' compensation and payroll taxes is a critical question requiring specialized legal advice.[86]

Under the federal Fair Labor Standards Act (FLSA), which establishes the federal minimum wage among other employment regulations, the United States Supreme Court has held that individuals performing work for an enterprise, even self-proclaimed volunteers that neither desired nor asked for compensation, are employees for the purposes of minimum wage.[87] The FLSA does exempt small

---

[80] Ibid.

[81] Ibid., 138.

[82] Lindsey Lusher Shute, *Building a Future With Farmers: Challenges Faced by Young, American Farmers and a National Strategy to Help Them Succeed* (2011): 23.

[83] Ibid.,141.

[84] Johnson & Endres, *Small Producers*, 110 n.38.

[85] Miller & Prizio, *Innovate Use*, 144.

[86] A. Bryan Endres and Rachel Armstrong, "Diverging Values: Community Supported Agriculture, Volunteers, and the Hegemonic Legal System" (working paper, University of Illinois, Department of Agricultural and Consumer Economics, 2012); A. Bryan Endres, Rachel Armstrong, and Megan Galey, "Worker Share Agreement Guide and Models" (Direct Farm Business: 2012).

[87] *Tony and Susan Alamo Foundation v. Sec'y of Labor,* 471 U.S. 290 (1985).

farms from complying with minimum wage rules.[88] However, for a diversified direct market operation commonly found in the local food system, the exemption is far from straightforward. The definition of agricultural labor, which is the basis of the small farm exemption, may not include such activities as working at a farmers' market stand or handling products grown on another farm.[89] As many CSAs distribute product from other farms, these distinctions have a significant financial impact on direct market operations.[90] On the other hand, many farms offer in-kind wages through farm products, thereby potentially meeting minimum wage obligations depending on the value of the produce. For example, allowing CSA members to work in exchange for a CSA share is quite common.[91] However, this allowance may be accompanied by payroll and other tax reporting complications—particularized topics that are not adequately addressed by the current Internal Revenue Service regulations and thereby injecting uncertainty into the farming operation.

Issues surrounding volunteers, worker shares and interns become especially complex when a farm considers the risk of injury. Workers' compensation is designed to address employee injuries, but many states exempt farmers from the requirement to purchase workers' compensation. However, many farmers do not realize that the standards for exemption from minimum wage and workers compensation may be different, or that the same complications in defining the scope of agricultural labor apply. In Illinois, small farms are exempt from minimum wage,[92] but the exemption from workers' compensation is narrower.[93] Further, workers' compensation provides protection with which many direct market farms may benefit; where workers' compensation insurance is available, an injured person may not sue the farm under a negligence theory.

The same tort liability issue for farm injuries relates to other types of farm visitors such as u-pick customers or individuals visiting roadside stands. In the past, Illinois law provided farms that offered agro tourism, such as farm tours or hiking trails, relief from negligence for failure to warn individuals if guests were injured while engaging in those activities.[94] However, after an Illinois Supreme

---

[88] *Code of Federal Regulations*, Employment in Agriculture That Is Exempted From the Minimum Wage and Overtime Pay Requirements Under Section 13(a)(6), Title 29, sec. 780.305;

[89] Amanda Heyman and Jennifer Jambor-Delgado, *Farmers' Guide to Farm Employees* (Farmers' Legal Action Group: 2012), Quick Q&A-2.

[90] A. Bryan Endres and Rachel Armstrong, "Diverging Values: Community Supported Agriculture, Volunteers, and the Hegemonic Legal System" (working paper, University of Illinois, Department of Agricultural and Consumer Economics, 2012)

[91] Ibid.

[92] 820 ILL. COMP. STAT. § 105/3(d)(2) (exempting agricultural enterprises employing less than 500 working days of agricultural labor per quarter in any quarter of the preceding year.)

[93] 820 ILL. COMP. STAT. § 305/3(19) (exempting agricultural enterprises employing less than 400 working days of agricultural labor per quarter in any quarter of the preceding year.)

[94] A. Bryan Endres and D. L. Uchtmann, "Survey of Illinois Law: The Latest Twist on the Illinois Recreational Use of Land and Water Areas Act: Clamping Down on Landowner Immunities" *Southern Illinois Law Journal* 29 (2004): 585-57.

Court decision[95] interpreted the Illinois Recreational Use of Land and Water Areas Act[96] to apply only to landowners who make recreational opportunities available to all of the public rather than to select individuals, the legislature amended the law. The amendments limited the protection from negligence liability only to those that engage in hunting or shooting based-recreational activities as opposed to traditional agro tourism .[97] The effect of the amendments to the Illinois Recreational Use Act has been to make agro tourism riskier for direct farm businesses offering recreation that is not hunting-related.[98]

Despite the plethora of legal issues arising from a direct farming operation—including food safety concerns—a survey of CSA's in Illinois and Wisconsin found that 87% had not contacted an attorney at any point in the business, leaving these operations vulnerable to all of the above mentioned issues.[99]

Advancing the Movement: Providing Legal Services to the Local Food System

Since the 1960's, the billable hour has been the dominant billing method used by non-contingency fee based firms, and is meant to ethically keep track of the time lawyers are working on client's issues.[100] The billable hour includes charges for travel, research, writing, and any other activity the lawyer does on behalf of the client.[101] Farmers often work odd hours and live in rural areas that may be far from an attorney's office. Thus under this traditional model, much of the farmer's money would end up paying for travel rather than legal services.[102] In addition, many farmers, especially beginning farmers, have the majority of their assets tied up in land costs and are unable to pay unpredictable, hourly rates for their complex legal issues.[103] Faced with the option of equipment repairs and maintenance or large legal fees to deal with issues the farmer likely perceives as distant or hypothetical, the farmer may well prioritize the mechanical running of the farm over compensating the lawyer to address any future and, hopefully, unrealized legal challenges.[104]

From the lawyer's perspective, representing complex operations such a direct farm business requires expertise in multiple regulatory subjects as opposed to the

---

[95] Hall v. Henn, 208 Ill.2d 325, 331 (2003).

[96] 745 ILL. COMP. STAT. 65/1-65/7 (2005).

[97] Ibid.

[98] A. Bryan Endres and D.L. Uchtmann, "New Recreational Use Act Rules For Illinois Landowner Liability: Two Steps Forward, One Step Back" *Agricultural Law and Taxation Briefs* 05-02 (2005): 1-2.

[99] A. Bryan Endres, Rachel Armstrong, and Megan Galey, "Legal Implications of Volunteer Work Opportunities on Community Supported Agriculture (CSA) Farms" (working paper, University of Illinois, Department of Agricultural and Consumer Economics, 2012); Johnson & Endres, *Small Producers*, 132.

[100] Susan Saab Fortney, "The Billable Hours Derby: Empirical Data on the Problems and Pressure Points," *Fordham Urban Law Journal* 33 (2005-2006): 171-72.

[101] Miller & Prizio, *Innovate Use*, 133.

[102] Ibid., 135-36.

[103] Miller & Prizio, *Innovate Use*, 133.

[104] Ibid., 133.

simplified and repetitive legal needs of most commodity cash-grain farms. Law offices that serve diversified farms must be prepared to address multiple issues, and not just specialize in a certain practice. On the other hand, seeking legal expertise at separate firms in a piecemeal fashion to address each subject matter is prohibitively expensive and may result in attorneys missing operational aspects that a lawyer working on all the farm's legal issues would otherwise identify.[105] Accordingly, lawyers that serve direct farm businesses will have to adjust their practices models order to maximize their ability to serve the direct farm business operation.[106] Although a few entrepreneurial attorneys have developed practice niches to serve the direct farm business client, for the most past farmers and attorneys are not connecting at the level necessary to sustain the local food movement.

The high cost of an attorney certainly presents one barrier to linking qualified attorneys with the entrepreneurial farmer. In Illinois, some legal aid offices, through the Illinois Family Farm Law Project, in collaboration with Farmers' Legal Action Group and the University of Illinois, provide direct attorney counsel to farmers on subjects such as business organization, credit, lender mediation, and disaster assistance.[107] As a result, low-resource entrepreneurs have access to the legal counsel they need to build long-term strength. However, affordability is just one barrier to legal services. In fact, farmers state that the primary and secondary reason that they do not seek legal advice is that attorneys do not understand their businesses or cannot help them.[108] These concerns are only amplified for the direct farm enterprise. Agricultural law as a field of study has been built up around agriculture as a commodity operation.[109] By moving away from a commodity structure, farms that sell to consumers are further challenged by the existing legal system.

Training a new generation of attorneys is a problem without an easy solution. Classes on small business agricultural law are still quite difficult to find, despite a growing number of student organizations illustrating keen student interest in this area of law.[110] Although law schools have recently pushed for more practical curricula, a large number of students graduate unprepared to launch the kind of practice that can skillfully serve the farm entrepreneur.[111] The problem is compounded by the already-complex legal needs of the direct farmer. Solutions to this intractable problem are only just beginning to emerge. The Illinois Bar

---

[105] Ibid., 132.
[106] Ibid., 136.
[107] "Illinois Family Farm Law Project - Home." Illinois Family Farm Law Project - Home. http://www.ilfamilyfarmlawproject.org/ (accessed November 29, 2012).
[108] Endres, et al. The Legal Needs of Farmers, 148-150.
[109] Susan A. Schneider, "A Reconsideration of Agricultural Law: A Call for the Law of Food, Farming, and Sustainability" *William & Mary Environmental Law & Policy Review* 34 (2009) 935.
[110] Susan Schneider, "Food & Agriculture: The New Trend In Higher Education " Agricultural Law (blog), August 31, 2008, http://aglaw.blogspot.com/2008/08/food-agriculture-new-trend-in-higher.html
[111] Steven C. Bennett "When Will Law School Change?" *Nebraska Law Review* 89 (2010): 94-97, 108-119.

Association has the beginnings of a section for lawyers involved with family farm representation. One nonprofit organization, Farm Commons, is planning a network of attorneys who share client services methodology, legal research, and collaborate on generating solutions. These programs show promise, but more must be done.

One approach is to enhance educational outreach activities to raise awareness in the direct farm business community of regulatory issues, thereby illustrating the value of professional legal services. Educating farmers about their legal relationships to workers, customers, suppliers, buyers, and landlords is an essential first step. When asked to self-assess their knowledge of legal issues, farmers give themselves mediocre rankings.[112] Evidence of what farmers are actually doing confirms this assessment accurate assessment. A recent survey by one of this article's authors found that the majority of farmers did not know if they had insurance for specific elements of their operation, were out of compliance with employment laws, and were unaware of several regulations that applied to their business.[113] Farmers' noncompliance with regulations and misunderstanding of their legal obligations is not because they are unwilling to learn. To the contrary, in a separate survey by the same author, the great majority of respondents were interested in attending an educational session on each of 15 named subjects, which ranged from seed saving to divorce to production contracts.[114]

Academic institutions play an important role in meeting the educational needs of direct market farmers who are creating new, innovative businesses that may not neatly align with existing law. In one illustrative situation, the legal aspects of the fast-expanding Community Supported Agriculture industry had gone unaddressed by farmers and attorneys alike. A University of Illinois research team tackled this problem by first surveying farmers for their experiences and subsequently investigated insurance, liability, contract and tax aspects of this unique arrangement between farmer and consumer. Going beyond research, the team developed actionable resources that farmers can access on-line and integrate into their business.[115] The development of model contracts, along with an accompanying guide, empowered farmers to identify and take action on these important issues. Online outreach and webinars complete the research team's activities—important mechanisms to extend the relevance of the academic

---

[112] A. Bryan Endres, Rachel Armstrong, and Megan Galey, "Legal Implications of Volunteer Work Opportunities on Community Supported Agriculture (CSA) Farms" (working paper, University of Illinois, Department of Agricultural and Consumer Economics, 2012).

[113] A. Bryan Endres, Rachel Armstrong, and Megan Galey, "Legal Implications of Volunteer Work Opportunities on Community Supported Agriculture (CSA) Farms" (working paper, University of Illinois, Department of Agricultural and Consumer Economics, 2012).

[114] A. Bryan Endres, et al. "The Legal Needs of Farmers: An Analysis of The Family Farm Legal Needs Survey." *Montana Law Review* 71 (2010): 147.

[115] "Direct Farm Business - Community Supported Agriculture." Direct Farm Business. http://www.directfarmbusiness.org/csa/ (accessed November 29, 2012).

research as farmers consistently rank the internet as their primary source of legal information.[116]

Other universities around the country are working to resolve the dearth of legal education available to farmers. In a complimentary approach, Drake University Agricultural Law Center has developed resources with a strong outreach focus, including online video content and farmer stories. Their comprehensive guides explore innovative land lease arrangements in an effort to address the predominance of leasing land by direct farm businesses.[117] In another example, Vermont Law School launched its Center for Agriculture and Food Systems in 2011 and has developed resource manuals for farmers.[118] The venerable Farmers Legal Action Group in Minnesota has produced legal manuals on the more complex legal issues vexing farmers, such as employment law and organic production contracts.[119] Because of these educational resources, farmers are better able to make crucial decisions and implement solutions.

Ultimately, however, the legal situation of the direct farmer is best addressed by meeting one on one with an attorney. Because the laws affecting these operations are many and complex, individual attention is the best way to protect each farm and foster community-wide growth. Additionally, the confidentiality of an attorney client relationship and the professional responsibilities of each attorney further protect the farmer. Training attorneys to better assist farmers and generate careful solutions to the complex range of problems faced by these farmers creates a comprehensive solution. Examples of these solutions point the way forward for the entire national food system.

Despite the legal and institutional barriers to the direct farm business model, the future is bright for their role in a more localized food system. But as the success of this business model grows, so too does its vulnerability to the legal morass of regulations, laws, and institutional barriers that may exempt the smallest operations. As regulatory relief for larger scale direct farm businesses is unlikely (and perhaps unwarranted), farmer and attorney education, as well as affordable legal service delivery models will be necessary to move the entrepreneurial direct market farmer towards a stable, secure future at the center of a revived local food system in Illinois and across the United States.

---

[116] Astrid Newenhouse and Sharon Lezberg, *Survey of Women Direct Market Farmers in Wisconsin* (Environmental Resources Center: 2009), 2; Sharon Lezberg and Julia Reyes-Hamann, *Hispanic Farmers in Wisconsin* (Environmental Resources Center: 2010), 4.

[117] "Sustainable Farm Lease: Home." Sustainable Farm Lease: Home. http://sustainablefarmlease.org/ (accessed November 29, 2012).

[118] "Center for Agriculture and Food Systems - Vermont Law School." Home - Vermont Law School.
http://www.vermontlaw.edu/Academics/Environmental Law Center/Institutes and Initiatives/Center_for_Agriculture_and_Food_Systems.htm (accessed November 29, 2012).

[119] Amanda Heyman and Jennifer Jambor-Delgado, *Farmers' Guide to Farm Employees* (Farmers' Legal Action Group: 2012); Amanda Heyman, *Farmers' Guide to Organic Contracts* (Farmers' Legal Action Group: 2012).

Chapter 8: Seeds and Sustainability: Why Keeping Seeds Local is an Act of Resistance and Resilience

Molly E. Hicks

What I'm aiming for is taking the local-food [SIC] movement to the next logical level, which is to establish a supply of locally grown seed as the underpinnings of a local food supply. Currently the seeds we order through commercial catalogs are grown literally all over the world. They are grown under radically different growing conditions than those that exist where we live. Further, we don't know how the seeds were grown, when they were harvested, or under what conditions they were processed, stored, or transported. It makes perfect sense to grow not just the plants but also the seeds in the area where the food will be consumed, giving them the opportunity to adapt on the deepest level.[1] – Sylvia Davatz, expert seed saver

Introduction

Over the past two decades, participating in the production, distribution, and consumption of local food has grown from a niche activity to a well-established movement that challenges the status quo of industrial agriculture. Here in the heartland, as in many other places in the United States, more and more small farms are making a profit through local food markets as the movement continues to expand and develop its methods and mission. One indispensable aspect of food production less considered by local food proponents is the place of seeds within local, sustainable agri-food networks (which are also called agro-food networks, community food systems, or local/regional food systems or networks). Until the advent of the commercial seed industry, most U.S. farmers and some home gardeners in the heartland and elsewhere saved seeds. It is still common in developing countries, though a majority of farmers in developed nations have

[1] Janisse Ray, *The Seed Underground: A Growing Revolution to Save Food* (White River Junction, VT: Chelsea Green, 2012), 69.

stopped this practice.[2] Saving seed for the next growing season increases individual and community food security by providing an assurance of future sustenance. Seed exchanges, where seed is given to or traded with others within and between communities, sustains and strengthens social relationships, and it provides a setting for the exchange of essential farming knowledge regarding seeds, plants, and farming techniques. Seed saving and exchanges also help to sustain crop diversity within different farming regions, which strengthens the genetic pool in local agri-food networks.[3] These practices increase the likelihood of successful future crops adapted to local environments.

As seed saving and the infrastructure that supported it decreased, the above-mentioned benefits were also lost. A growing body of literature addresses how the industrialization and commercialization of seeds has adversely affected both natural and social systems. As stated by the International Commission on the Future of Food and Agriculture, "The disappearance of local seeds has gone hand in hand with the disappearance of small farmers and local food cultures. And with them, local knowledge about the use of cultivated and wild plant varieties in their different ecological and cultural habitats has likewise been lost."[4] An estimated 75 percent of crop diversity has been lost globally since the beginning of the 20th century.[5] These seed extinctions have occurred in conjunction with the growth of the commercial seed industry and industrial agriculture, and the degradation of the natural and social systems within which these industries operate. Yet, even though seeds are the most essential part of agricultural production, local food proponents do not consistently focus on the localization of seed stocks to further counter the influence of the industrial regime. In the heartland and other parts of the United States, farmers and gardeners still purchase the majority of their seed from outside their growing region. Local seed saving and community seed exchanges can provide both resistance to industrial agriculture, and another way to increase resilience in local agri-food networks.

This chapter will explore why locally produced and locally saved seed is an important aspect of resilience in local agri-food networks. Common goals in the local food movement will be analyzed to assess whether they address the localization of seeds in any way. From this assessment, an argument will be built in support of the development of regional seed banks and libraries, and seed saving and exchange programs, which are essential for the success of sustainable agri-food networks. But first, it is important to understand how seeds transitioned from a community-owned natural resource to a commodity controlled by corporate interests. This will be done through a brief historical exploration of seeds in the United States.

---

[2] Abby Kinchy, *Seeds, Science, and Struggle*: *The Global Politics of Transgenic Crops* (Cambridge, MA: MIT Press, 2012), 104.

[3] Vandana Shiva, ed., "Manifesto on the Future of Seed," in *Manifestos on the Future of Food and Seed* (Cambridge, MA: South End Press, 2007), 77-79.

[4] Ibid., 80.

[5] Ibid., 80 (footnote).

A Brief History of Seed in the Heartland and Beyond

Until a government program that distributed free new seed varietals from overseas gained large support in the 19th century, farmers throughout the Midwest and in other parts of the United States were saving their own seed.[6] Immigrants coming to the United States would often bring seeds with them to plant in their new gardens.[7] Some slaves also brought seeds with them, which is how okra, peanuts, cowpeas ("black-eyed peas"), and sesame arrived in the United States.[8] The early commercial seed industry initially catered to home gardeners. This new industry first became successful in colonial New England, and it grew steadily as settlers moved westward into the Midwest.[9] By the mid-19th century, the industry was supplying seed to a growing population of landless home gardeners living in urban areas, and a small but significant group of commercial produce growers.[10] By the 20th century, the aforementioned government seed program, commercial seed sales, and industrial agriculture had greatly reduced local seed saving and exchange in the United States. The majority of the heartland's economy became focused on commercial livestock and hybrid varieties of feed grains, such as corn north of the Ohio River and wheat in the northern Midwest.[11] Fruit and vegetable seed markets grew as well during this time. Most farmers stopped saving seed and began relying upon the purchase of hybrid seeds. These newer varieties would reliably produce in the first year higher yields, called "hybrid vigor."[12] American farmers and home gardeners had been convinced to purchase what at one time was a free and easily obtained agricultural resource.

Hybrid crop breeding programs, which first developed at land-grant universities around 1890, created high-yielding crop varieties grown to withstand specific pests and diseases. Starting around 1935, the meaning of "hybrid" changed from a simple cross between two naturally occurring varieties to a cross between two already inbred varieties.[13] Genetically speaking, these varieties are called F_1 hybrids. Further hybrid lines called F_2, F_3, and so forth can be created as well. Hybrids became a lucrative business in the mid-20th century. (Fig. 1) During what is called the Green Revolution, farmers across the United States and abroad quickly became reliant upon hybrid seeds developed through university and private seed company research. These new hybrids required chemical inputs in the form of fertilizers, herbicides, and pesticides, which led to reliance upon the purchase of these production materials along with seeds in order to enhance plant

---

6 Jack Ralph Kloppenburg Jr., *First the Seed: The Political Economy of Plant Biotechnology*, 2nd ed. (Madison: University of Wisconsin Press, 2004), 61.
7 Steve A. Eberhart and Lauren A. Wiesner, "Long-term Storage of Rye at the National Seed Storage Laboratory" (presentation, International Conference on Crop Germplasm Conservation with Special Emphasis on Rye, Warsaw/Konstancin-Jeziorna, Poland, July 2-6, 1996).
8 Mary Tolford Wilson, "Peaceful Integration: The Owner's Adoption of His Slaves' Food," *Journal of Negro History* 49, no. 2 (1964): 116.
9 Ulysses Prentiss Hedrick, *A History of Horticulture in America to 1860*, 2nd ed. (Portland, OR: Timber Press, 1988), 40-42.
10 Kloppenburg, *First the Seed*, 61.
11 Ibid., 57-61.
12 Ray, *The Seed Underground*, 11.
13 Kloppenburg, *First the Seed*, 68.

growth and yields.[14] As Raj Patel notes, "In order to work, the seeds required almost laboratory-perfect growing conditions, which demanded irrigation, fertilizers and pesticides. These, in their turn, depended on fossil fuels for their production. And the entire enterprise required expunging of native biodiversity, so that rows of the new seed might take its place."[15] Green Revolution hybrids and chemical inputs have created massive environmental deterioration within agricultural systems through the deterioration of soil and water quality, and an increase in ecological imbalances between insects, animals, soil microbes, and other plants present within natural systems that surround farmland.

Fig. 1. Hybrid seed variety, advertised in Oscar H. Will & Co. catalog, 1951.[16]

Pest and disease resistance is fostered in ecological systems through natural selection. Likewise, seed savers also foster hardiness, vigor, and a better genetic fit between the plants they raise and the local ecosystem. They do this by saving seeds from their strongest and healthiest plants that exhibit important characteristics they want to pass on. The crop plant genetic pool has been eroded through intensive hybrid seed development that uses only a small number of plant varieties. This has greatly reduced crop diversity and natural resistance to local environments once managed through local seed saving and exchange. We have essentially "built our roof with stones from the foundation."[17] Natural forms of resistance, adaptation, and resilience were ensured by an abundant heterogeneous gene pool of seed varieties. Seeds were managed and saved by humans selecting for plant diversity and desired traits, which inherently built resilience into

---

[14] Cary Fowler and Pat Mooney, *Shattering: Food, Politics, and the Loss of Genetic Diversity* (Tucson: University of Arizona Press, 1990), 47-48, 55-63.
[15] Raj Patel, *Stuffed and Starved: The Hidden Battle for the World Food System* (Brooklyn, NY: Melville House Publishing, 2007), 120.
[16] Fred Schneider, "'Corn in the Crib is Like Money in the Bank': George F. Will and the Oscar H. Will & Company, 1917-1955," *North Dakota History: Journal of the Northern Plains* 76, no. 1&2 (2010): 17.
[17] Ibid., xii.

agricultural systems. This varied gene pool is now being replaced at an alarming rate by the persistently homogenizing field of agricultural science.

Further corporate control was achieved in the 1970s and 80s through the passing of intellectual property rights laws such as plant breeders' rights and patents. As Abby Kinchy notes, "Since the 1980s, the international governance of plant genetic resources has shifted from a common heritage system, which treated plant genetic resources as a commons, to a system that emphasizes sovereign and private property rights."[18] Biotechnology patents soon followed. These patents allow a seed, plant, or organism that has been genetically modified to be owned by the creator, which may be an individual or a corporation. Even if farmers could save hybrid seed, which is not very successful since seed grown the second year from a hybrid plant does not tend to grow true-to-type, they are legally barred from doing so because of intellectual property rights laws. This is true for genetically modified varieties as well, which may or may not be hybrids. Additionally, some modified seeds contain genetic use restriction technology, called "terminator seeds" by anti-biotechnology activists. They are genetically altered so that plants grow sterile seeds, which limit their use to one growing season.[19] Resistance to these changes in seed production, and other socially and environmentally unsustainable aspects of industrial agriculture, led farmers, gardeners, and activists to develop the local food movement in the late 20th century.

Why Save Seed?

Seed saving occurs at many levels. There are individual gardeners who save seed, community gardens or other groups of seed savers that run local seed banks or seed libraries, and non-profit organizations and commercial businesses that grow out and save seed to be sold, donated, or preserved. Larger-scale seed saving entities include the National Center for Genetic Resources Preservation in Ft. Collins, Colorado, which saves seed, plant, and animal genetic resources that comprise the United States' agricultural economy. The Svalbard Global Seed Vault located on the Norwegian island of Spitsbergen provides long-term storage for the world's food crops to ensure sustenance for future generations. There may be as many reasons to save seed as there are ways and scales at which to do so. The following provides both social and ecological reasons why organic, open-pollinated seed saving is such an important endeavor.

To Reap the Benefits of Open Pollination

To understand the importance of saving non-hybrid seed, one must understand a little about what the process of open-pollination entails. Open pollination means plants are able to freely pollinate either between flowers on a single plant (self-pollination or inbreeding) or between flowers on different plants (cross-pollination or outbreeding). All heirloom seeds passed down from generation to

---

[18] Kinchy, *Seeds, Science, and Struggle*, 104.
[19] Ibid., 107.

generation are open-pollinated varieties.[20] Seeds that grow from plants that have been openly pollinated will have greater genetic diversity, which, as will be explained later, helps increase resilience in crops. Open pollination is the opposite of controlled pollination, where two parent strains with known traits are crossed in controlled conditions to create a hybrid variety. Controlled pollination (done by humans) and cross-pollination (done either by wind, insects, birds, or humans) both create hybrid seeds if the crosses are between two plants of different varieties in the same species. This may not be a negative situation if one is looking to find a new crop to stabilize over a certain number of growing seasons. In fact, many heirlooms available today began through intentional or mistaken crosses between different varieties. Seeds were saved from these "natural" hybrids, and over many growing seasons they become a stable new variety. As mentioned previously, this cannot happen with an F_1 hybrid seed created through controlled pollination because it has been crossed with two inbred lines, and the seeds will not grow plants that are true-to-type in the coming year. Allowing open-pollination of seed, while also making sure that plants aren't crossing with different varieties of the same species unless one wants them to (one of the challenges of seed saving that requires attention to detail), will result in a stock of seeds that over time are better adapted to local pests, diseases, and environmental stressors like temperature fluctuations and extreme weather events.

To Increase Food Security and Re-establish Community Control

In a world where corporate interests, climate change, and environmental degradation continue to negatively affect agriculture, food access, and food prices, seed saving can provide further assurance to local farmers and home gardeners within the local food movement that they will be able to grow plants adapted to their local environment year after year. As Kloppenberg states, "If true food sovereignty is to be achieved, control over genetic resources must be wrested from the corporations and governments that seek to monopolize them and be restored to, and permanently vested in, social groups and/or institutions with the mandate to sustain them and to facilitate their equitable use."[21] Seed saving is an act of resilience because it increases agricultural sustainability by using locally adapted seeds that over time can better withstand climate change and other environmental stressors. It is also an act of resistance to the commercialization and industrialization of an essential component of agriculture that instead belongs in the hands of those who produce and eat its bounty.

---

[20] For more information on pollination and seed saving, see Suzanne Ashworth, *Seed to Seed: Seed Saving and Growing Techniques for the Vegetable Gardener,* 2nd ed. (Decorah, IA: Seed Savers Exchange, 2002).

[21] Jack Ralph Kloppenburg Jr., "Seed Sovereignty: The Promise of Open Source Biology," in *Food Sovereignty: Reconnecting Food, Nature and Community*, ed. Hannah Whittman et al. (Oakland, CA: Food First Books, 2010), 152.

To Increase Genetic Diversity

As mentioned in the introduction, an estimated 75 percent of crop diversity has been lost globally since the early 20th century due to unsustainable agricultural practices. This has led to a massive decrease in genetic diversity in crop plants, and an increase in the risk of disease and pest infestations. Fowler and Mooney explain, "As agriculture took hold and developed in ancient times...a certain fragile balance came to exist between plants, pests, and diseases. Primitive varieties – landraces – exhibited a great deal of genetic variation...to be sure, pests and diseases struck, but their attacks were muted by the diversity and strength of the defenses accumulated by the plants during thousands of years of adaptation under agriculture."[22] These adaptations were cultivated through the careful selection of seed from plants that remained healthiest after pest or disease attacks. The modern use of pesticides, fungicides, herbicides, and the genetic modification of seed to resist pests and disease, has yet to win a seemingly endless battle. Both insects and diseases adapt to chemicals: they can mutate and overcome resistance that has been genetically bred into plants.[23]

With regard to seed saving, Carine Pionetti notes, "The continuous exchange of seeds for local crop varieties circulates genetic resources from one field to another...The dynamic management of genetic resources enhances the stability of traditional [and modern alternative] agrosystems, increases the adaptation potential of local crops to evolving environmental conditions and limits the risk of genetic erosion."[24] The best way to increase resilience and address the never-ending presence of pests and diseases is to reestablish genetic diversity within local seed stocks and farming systems. Planting diverse types of open-pollinated, saved seeds in biodiverse farming systems as opposed to hybrid monocropping, which increases the speed of pest and disease adaptation, builds crop resistance through genetic diversity.

To Save Family and Community Traditions

Saving seeds allows gardeners and farmers to take part in history by passing on knowledge not only through the genetics of the seed, but also through the familial and cultural stories told by those who have saved seed from year-to-year. Along with increasing genetic diversity and saving familial and cultural heritage, seed saving can encourage the passing of other knowledge associated with crop production and food processing, including traditional gardening and backyard harvesting techniques, and methods for traditional food storage and preparation.[25] Suzanne Ashworth asserts, "Untold numbers of old-time varieties are lost each year, because elderly gardeners can no longer find family members willing to grow and maintain these living heirlooms. When elderly seed savers pass away,

---

[22] Fowler and Mooney, *Shattering*, 42

[23] Ibid., 49.

[24] Carine Pionetti, *Sowing Autonomy: Gender and Seed Politics in Semi-Arid India* (London: International Institute for Environment and Development, 2005), 154.

[25] Gary Paul Nabhan, *Renewing America's Food Traditions: Saving and Savoring the Continent's Most Endangered Foods* (White River Junction, VT: Chelsea Green Publishing, 2008), 99.

unless their seeds are replanted by other gardeners, their outstanding strains become extinct. Invaluable genetic characteristics are lost forever to future generations of gardeners and plant breeders."[26] Preserving an aspect of local heritage through seed saving, and passing these seeds to others to grow, strengthens communities through knowledge sharing.

Seed saving is a challenging endeavor for those with little time, like many small-scale farmers in the local food movement who are working long hours to simply break even each year. Their commitment to local, sustainable agriculture is strong, but it does not necessarily mean all have the ability and time to integrate seed saving into their farming practices. The growth of seed banks and libraries (which will be discussed later in this chapter) is turning seed saving back into a community practice organized and managed collectively by community members and non-profit organizations. Regional seed banks and libraries could provide locally adapted seed not only to home and community gardeners, but to farmers as well. This ultimately increases sustainability within the local food movement through supporting local seed saving, seed exchange, and seed distribution to farmers and gardeners who provide sustenance for the region where they live and work.

To Keep Things Local

Saving seed supports the movement to localize food production, distribution, and consumption. Many aspects of agriculture, food processing, and consumption are being localized through work being done in the local food movement. This includes the localization of funding streams that support small farms and local food initiatives, localizing producer-distributor and producer-consumer links, farm-to-table events, farmer cooperatives, community-owned food cooperatives, and community gardens that support neighborhoods, immigrant communities, and food banks. At the core of all of this work is the seed, which grows plants that provide food for both humans and livestock. Are those within the local food movement cognizant of the connections between local seed production, sustainability, and resilience within agri-food networks? To address this, the following section will investigate the mission statements and activities of two well-known organizations that promote the advancement of a local food agenda in order to uncover how seeds are, or are not, included. The two organizations are Sustainable Agriculture Research and Education Program, which conducts applied research and education from an academic institution, and Slow Food International, a non-profit local food organization.

---

[26] Ashworth, *Seed to Seed*, 13.

Seed and the Local Food Movement

Sustainable Agriculture Research and Education Program

Over the past 15 years, Gail Feenstra has become a well-known local food scholar and researcher within the movement. Feenstra and her colleagues at University of California's Sustainable Agriculture Research & Education Program (SAREP) define a community food system as, "a collaborative network that integrates sustainable food production, processing, distribution, consumption and waste management in order to enhance the environmental, economic and social health of a particular place. Farmers, consumers and communities partner to create a more locally based, self-reliant food economy."[27] The seven stated goals of SAREP's community food system are as follows:

1. A stable base of family farms that use sustainable production practices and emphasizes local inputs
2. Marketing and processing practices that create more direct links between farmers and consumers
3. Improved access by all community members to an adequate, affordable, nutritious diet
4. Food and agriculture-related businesses that create jobs and recirculate financial capital within the community
5. Improved living and working conditions for farm and food system labor
6. Creation of food and agriculture policies that promote local or sustainable food production, processing and consumption
7. Adoption of dietary behaviors that reflect concern about individual, environmental and community health[28]

These seven goals reflect common elements found in local food mission statements that have been developed by various academic institutions, departments, and programs, non-profit organizations, government entities, restaurants and other food institutions, community groups, and individuals. Research conducted by SAREP that address these seven goals include farm-to-school and farm-to-hospital studies, agriculture and marketing, energy use and food production, agriculture and water management, and sustainable sourcing of raw materials.[29] So far, no research on seeds appears to have been conducted through SAREP, and there is little focus on the importance of local seed sources in its educational programs. Educational programs and presentations conducted about seed saving are geared toward home gardeners. Seed saving is being treated solely as a garden hobby (and a fine hobby it is!) instead of also as an essential part of food production within the local food movement.

---

[27] Sustainable Agriculture Research and Education Program, *Defining Sustainable Community Food Systems*, http://www.sarep.ucdavis.edu/sarep/sfs/def.

[28] Ibid.

[29] Sustainable Agriculture Research and Education Program, *ASI Research*, http://www.sarep .ucdavis.edu/research.

The first and fourth goals in SAREP's vision of a community food system have the potential to address the localization of seeds. First, sustainable production practices that use local inputs (goal 1) might include the seed through community organized seed saving among home and community gardeners, and farmers that are willing and able to save seed. Seed libraries in the United States have grown in number and size in the past decade. Either community-organized or run by local non-profits, these libraries "lend" free seed to gardeners and farmers with an agreement that some of this seed be grown out, saved, and returned after harvest time. Some are in actual libraries that have been able to set aside space for this venture. Pima County Seed Library in the Tucson, Arizona region is managed in part by the library system, and seed is housed in six of the county branches. There is an online seed catalog that interested farmers and gardeners can browse, and seed is checked out with a library card.[30] Seed banks are similar in their organization, but they may have a pricing structure in place for the purchase of saved seed. These banks and libraries have the potential to become non-profit entities in and of themselves, which allows greater access to private and federal funding to stabilize their presence within communities. This could lead to employment opportunities for local community members (goal 4), which further supports the economic sustainability aspect of the local food mission.

Seed Savers Exchange (SSE) in Decorah, Iowa is a non-profit seed saving organization. It has an extensive seed preservation bank, and it also sells seed through a commercial catalog. SSE employs over 60 local residents who grow out, process, package, save, and sell seed to customers across North America and abroad. Other employees work in education and programming, development, human resources, seed genetic resource preservation, plant collections, the gift shop, and facilities maintenance. The organization has grown from a handful of employees to over 60 in less than a decade because of the immense growth of home and community organic gardening. SSE's organizational model could be used at the regional scale to provide local seed and local income for residents in future community food systems like the one described by SAREP.

Hudson Valley Seed Library in upstate New York is a potential model for future seed companies focused on both regional and organic seed development. It began as a seed lending library in 2004, and today it is a company dedicated to growing out and saving organic, non-hybrid seeds adapted to the region of upstate New York where the seed farm is located. Profits come from a seed catalog that sells both seed grown on the small farm and varieties purchased from other seed houses dedicated to the same ethical and environmentally sustainable principles.[31] The owners also encourage seed library members living in the region, whose membership is another source of income, to take part in saving certain varieties each year, which are then donated to community gardens, food banks, and school gardens. Members receive education and tips on saving seed, informational emails, and discounts on seed.[32] This model is a promising one for the future of

---

[30] Pima County Public Library, *Seed Library*, www.library.pima.gov/seed-library/#return.

[31] Hudson Valley Seed Library, *Hudson Valley Seed Library*, http://www.seedlibrary.org/about-us-hvsl/.

[32] Hudson Valley Seed Library, *Membership*, http://www.seedlibrary.org/membership/.

commercial seed in the local food movement. Nonetheless, shifting from a commercial model to one that allows for community ownership of seed will further what should be a central goal in the local food movement of placing seed back into the environments where its crops are both produced and eaten, and where it can be freely saved, exchanged, and used *ad infinitum*.

SAREP's seven community food system goals reveal the complexity involved in growing a local food economy. Scholars and practitioners working within the program have developed relevant and timely research that furthers these goals. What is largely missing from this scholarship, and is only found piecemeal within educational programming, are conversations about the fundamental importance of seed in local agri-food network development. There is a larger discourse missing about how this essential component of agriculture should have a localized place within agri-food networks to better reflect the sustainability ethic that is being encouraged by local food proponents.

Slow Food International

Slow Food International, which has a U.S. office in New York City and various chapters throughout the country, is one of the oldest and possibly most well-known local food organizations. Beginning in Italy in 1989, Slow Food (which is another term for local food) envisions "a world in which all people can access and enjoy food that is good for them, good for those who grow it and good for the planet." The mission for the organization is to promote "good, clean, and fair food for all," which translates to food that is of good quality, food that is not harmful to people, animals, or the planet, and food that is accessible in price for consumers, and provides fair pay and working conditions for small-scale farmers and other producers.[33] Within their mission description, Slow Food indirectly discusses seeds through their commitment to "defending the biodiversity of cultivated and wild varieties...[and] maintaining the diversity of regional food and agricultural traditions."[34] For those who know that the only way to preserve traditional cultivated plants is to save seed from them, it is easy to see the connections between seed, biodiversity, and traditional farming. For others who are not aware of this connection, the seed maintains a status of less importance within the overall mission.

The seed is addressed within Slow Food's food sovereignty and anti-GMO (genetically modified organism) campaigns, and the Ark of Taste program. In the GMO campaign, the seed is discussed in light of the organization's aim to move agriculture away from its growing reliance on genetically modified plants and seeds, and instead refocus efforts on those produced sustainably, and selected and saved by farmers freely on their own land.[35] Within the food sovereignty campaign, seeds are mentioned briefly in the campaign description when describing the Commission on the Future of Food and Agriculture's "Manifesto

---

[33] Slow Food International, *Our Philosophy*, http://www.slowfood.com/international/2/our-philosophy.
[34] Ibid.
[35] Slow Food International, *GMOs*, http://www.slowfood.com/international/22/gmos.

on the Future of Seed."[36] Slow Food's leader, Carlo Petrini, was a part of creating this international manifesto in 2006 (along with "Manifesto on the Future of Food" in 2004 and "Manifesto on Climate Change and the Future of Food Security" in 2008). Slow Food worked with various other organizations and individuals within the local food movement, including North American local food leaders, academics, and activists Wendell Berry, Raj Patel, Frances Moore Lappé, Alice Waters, Miguel Altieri, and Percy Schmeiser.

The seed manifesto is an extremely impactful message against the industrialization and commercialization of seeds that includes "a new paradigm for seed." This paradigm, which fits well into Slow Food's overall local food mission, envisions a future without corporate seeds, large industrial farms, monocultures, and the resultant loss of natural resources, culture, traditional knowledge, and plant biodiversity.[37] Under the manifesto's section, "Living Alternatives – Seeds of Hope," one will find detailed mention of the recent growth of seed initiatives across the world. These include the growth of seed banks to preserve plant diversity, the increasing (and also ancient) role of women as seed savers and protectors in many farming communities, seed saving and exchange initiatives spurred by the growth of home and community gardening in both developing and developed countries, plant breeding projects focused on developing organic seed varieties for use in ecological agriculture, work to establish GMO-free zones, and alliances between groups in civil society working to pass legislation that encourages the development of local seed initiatives.[38]

The U.S. Ark of Taste program, which is an international program for Slow Food as well, works to save knowledge about culturally important foodstuffs including breads, cheeses, beverages, and the seeds of grains, fruits, and vegetables. This furthers the mission of the organization through the preservation of food and plant diversity. Slow Food USA briefly discusses seed saving and seed exchange as a way to become involved in saving different foods that are at risk of extinction. Unfortunately, there are no resources available along with this commentary to help those interested in seed saving understand how to start this endeavor.[39] Like with many other local food organizations, saving seeds is encouraged, but there is little education provided to help gardeners and farmers develop this sustainable farming practice and understand the deeper implications of saving seed.

Seed Saving Organizations

Organizations within the United States that focus specifically on non-industrial, non-hybrid seeds and seed saving include Seed Savers Exchange in Decorah, Iowa, Native Seeds/SEARCH in Tucson, Arizona, Organic Seed Alliance in Port Townsend, Washington, Occidental Arts and Ecology Center in Occidental,

---

[36] Slow Food International, *Food Sovereignty*, http://www.slowfood.com/international/16/food-sovereignty.
[37] Shiva, "*Manifesto on the Future of Seed*," 89-91.
[38] Ibid., 100-101.
[39] Slow Food USA, *Ark of Taste*, http://www.slowfoodusa.org/index.php/programs/details/ark_of_taste/.

California, and Seed Matters, a special initiative funded by the Clif Bar Family Foundation. (In 2009, the foundation pledged $1 million to "protect and improve living seed" to increase the health of local food systems.)[40] Almost all of these organizations share a seed saving and seed exchange mission among home and community gardeners, with a secondary interest in encouraging farmers to save seed. Organic Seed Alliance and Seed Matters specifically focus on advancing organic seed production and the development locally adapted seeds to increase sustainability in agri-food networks.

Even within these organizations, one will find a spectrum of discussions with regard to how seed should be considered in the local food movement, from a focus on seed saving with home gardeners, to building community seed exchange programs, to developing a more bioregional approach to the localization of seed stock. Organic Seed Alliance presents a convincing argument in its proposal of regional organic seed systems.[41] This idea warrants more discussion within the local food movement in conjunction with other initiatives like the development of regional "food hubs" to support the marketing and distribution of local foods,[42] or bioregional/ecoregional foodsheds that focus intentionally on the ecological as well as social health of local food networks.[43]

Re-establishing a Place for Seed

Dillon and Hubbard, of Organic Seed Alliance, propose the creation of regional organic seed systems in order to better meet the needs of organic farmers who often struggle to find enough seed to meet their demands. Their 2011 report, "State of Organic Seed," is a review of data about organic seed in the United States collected from surveys and questionnaires distributed to farmers, certifiers, researchers, experts in food and farm policy, and experts within the food and seed industries. The authors argue for the development of regional seed systems to support local farmers. They assert, "Organic seed that is appropriate for regional agronomic challenges, market needs, regulations, and the social and ecological values of organic agriculture is therefore fundamental to the success of organic farmers and the food system they supply."[44] What these regional seed systems will actually look like has yet to be uncovered, but this report is the first step in identifying what is needed to move more intentionally toward the localization of seed within agri-food networks in the United States. Models for development within a Western nation might be found by studying grassroots European seed networks that have developed in the past 15 years in countries including Spain, Italy, France, Scotland, Germany, Switzerland, Portugal, Bulgaria, and Hungary. For example, in Spain the six-family farming cooperative, La Verde, produces a

---

[40] Clif Bar Family Foundation, *Special Initiatives*, http://clifbarfamilyfoundation.org/Special-Initiatives.

[41] Dillon and Hubbard, "State of Organic Seed," v.

[42] United States Department of Agriculture, *Farmers Markets and Local Food Marketing*, http://www.ams .usda.gov/AMSv1.0/foodhubs.

[43] Gary Paul Nabhan, "Resilient Foodsheds: Agro-ecoregional Approaches to Food Security" (presentation at the National Bioneers Conference, San Rafael, California, October 19-21, 2012).

[44] Dillon and Hubbard, "State of Organic Seed," v.

year-round supply of organic fruits and vegetables to the region in which they operate. Due to a lack of commercial varieties that meet the needs of the organic growers, the cooperative has developed the largest farm-saved seed bank in Spain, which supplies the majority of small-scale organic farmers in the region.[45]

Regional seed systems should be focused on the production of seed for local farmers as well as the saving of seed to increase sustainability within agriculture and gardening in local agri-food networks. Matthew Dillon has been an avid seed saver and seed saving proponent for years.[46] He surely would agree to this addition, though it is not the focus of this organic seed report. Within these proposed seed systems, there must also be consideration of the importance of genetic diversity within seed stock, which can occur through the collection of numerous varieties of each type of crop. Will Bonsall, a seed saver who runs Scatterseed Project in Maine, has over seven hundred varieties of potato alone.[47] This is just one example of the crop diversity available to gardeners and farmers in the United States. There is ample opportunity to increase genetic diversity, and ultimately resilience, in local crops through the introduction of new varieties and the saving of seed from the healthiest, most adaptable plants from these varieties. Those in the local food movement must be in the business of what Vandana Shiva calls "cultivating diversity."[48]

This chapter has identified a need for local seed to supply farmers as well as home and community gardeners. A regional seed system, or seed network, which may be a better term for what is actually being created, should be designed to include all individuals and groups involved in locally marketed crop cultivation within that region's agri-food network. Not everyone will save seed, but everyone can reap the benefits from those who do – those who take part in the mission to preserve and enhance genetic and crop diversity in their local foodstuffs. There are already seed savers in every state. Many are listed in the annual Seed Savers Exchange Yearbook. Members can peruse seed offerings that have been saved by home gardeners across the United States and abroad, and request some via phone, mail, or email. Organizing local seed savers in a similar fashion will be an important part of the development of regional seed networks.

If standing in summertime at the highest point in Urbana, Illinois – the central flatlands of the state – one will see outside of the city a sea of hybrid corn fields planted from seed owned by corporations like Monsanto, Syngenta, and DuPont under brand names such as Dekalb, Pioneer, and Agrisure. None of these farms reflect principles within the local food movement, but sandwiched in between the acres and acres of industrial farmland are small farms of the future. Residents can purchase produce, dairy, eggs, and meat from these farms at local farmer's markets, the local food coop, and the natural food store. They can taste

---

[45] Elisa Da Vià, "Seed Diversity, Farmers' Rights, and the Politics of Re-peasantization," *International Journal of Sociology of Agriculture & Food* 19, no. 2 (2012): 234-235.

[46] Matthew Dillon, "Stewarding Seeds: Preserving Our Plant Genetic Resources" (presentation at the National Bioneers Conference, San Rafael, California, October 19-21, 2012).

[47] Ray, *Seed Underground*, 64.

[48] Vandana Shiva, *Stolen Harvest: The Hijacking of the Global Food Supply* (Cambridge, MA: South End Press, 2000), 112.

creations made from these products at area restaurants that buy locally. These farms are organically certified or managed without the use of pesticides and other harmful chemicals, and their caretakers are dedicated to environmental stewardship, energy efficiency, and humane treatment of animals. Statistically, few, if any of these farms, or other farms like them across the United States, save seed or have access to local seed banks or libraries. All of them have the potential to increase the sustainability of their operations, and the sustainability of the local food movement they are a part of, if they were able to save and/or access locally-grown seed through regional seed networks.

Starting with locally adapted seeds and encouraging genetic diversity through the cultivation of a diverse seed stock should be the foundation of any local food movement. From this starting point, the movement can develop stronger mechanisms to increase overall sustainability in agri-food networks. Regional seed stock saved and managed by seed libraries or banks will put local seeds into the hands of farmers, teachers and children in school gardens, community gardeners, institutional gardeners in hospitals and prisons, and home gardeners. The food produced from these seeds will feed the farmers and gardeners, their families, their customers, and those they give to freely. The seed bounty that is also produced will ensure the continuation of this diverse and nutritious bounty for the coming year. If a pest outbreak occurs, or a mid-summer drought, seed from those plants that survive best will be saved. In this manner, with every passing year, this seed network can support its regional agri-food network by providing plants that are more resilient and continuously adapting to local environments and the unpredictable future ahead.

References

Ashworth, Suzanne. Seed to Seed: Seed Saving and Growing Techniques for the Vegetable Gardener, 2nd ed. Decorah, IA: Seed Savers Exchange, 2002.

Clif Bar Family Foundation. Special Initiatives. http://clifbarfamilyfoundation.org/Special-Initiatives.

Da Vià, Elisa. "Seed Diversity, Farmers' Rights, and the Politics of Re-peasantization," International Journal of Sociology of Agriculture & Food 19, no. 2 (2012): 229-242.

Dillon, Matthew. "Stewarding Seeds: Preserving Our Plant Genetic Resources." Presentation at the National Bioneers Conference, San Rafael, California, October 19-21, 2012.

Dillon, Matthew and Kristina Hubbard. "State of Organic Seed." Organic Seed Alliance. https://www.seedalliance.org/Publications/#SOS, 2011.

Eberhart, Steve A. and Lauren A. Wiesner. "Long-term Storage of Rye at the National Seed Storage

Laboratory." Presentation at the International Conference on Crop Germplasm Conservation with Special Emphasis on Rye, Warsaw/Konstancin-Jeziorna, Poland, July 2-6, 1996.

Fowler, Cary and Pat Mooney. Shattering: Food, Politics, and the Loss of Genetic Diversity. Tucson, AZ: University of Arizona Press, 1990.

Hedrick, Ulysses Prentiss. A History of Horticulture in America to 1860. 2nd ed. Portland, OR: Timber Press, 1988.

Hudson Valley Seed Library. Hudson Valley Seed Library. http://www.seedlibrary.org/about-us-hvsl/.

--. Membership. http://www.seedlibrary.org/membership/.

Kloppenburg Jr., Jack Ralph. First the Seed: The Political Economy of Plant Biotechnology, 2nd ed. Madison: University of Wisconsin Press, 2004.

--. "Seed Sovereignty: The Promise of Open Source Biology." In Food Sovereignty: Reconnecting Food, Nature and Community, ed. Hannah Whittman et al. (Oakland, CA: Food First Books, 2010).

Kinchy, Abby. Seeds, Science, and Struggle: The Global Politics of Transgenic Crops. Cambridge, MA: MIT Press, 2012.

Nabhan, Gary Paul. Renewing America's Food Traditions: Saving and Savoring the Continent's Most Endangered Foods. White River Junction, VT: Chelsea Green, 2008.

--. "Resilient Foodsheds: Agro-ecoregional Approaches to Food Security." Presentation at the National Bioneers Conference, San Rafael, California, October 19-21, 2012.

Patel, Raj. Stuffed and Starved: The Hidden Battle for the World Food System. Brooklyn, NY: Melville House Publishing, 2007.

Pima County Public Library. Seed Library. http://www.library.pima.gov/seed-library/#return.

Pionetti, Carine. Sowing Autonomy: Gender and Seed Politics in Semi-Arid India. London: International Institute for Environment and Development, 2005.

Ray, Janisse. The Seed Underground: A Growing Revolution to Save Food. White River Junction, VT: Chelsea Green, 2012.

Schneider, Fred. "'Corn in the Crib is Like Money in the Bank': George F. Will and the Oscar H. Will & Company, 1917-1955." North Dakota History: Journal of the Northern Plains 76, no. 1&2 (2010): 2-25.

Shiva, Vandana, ed. "Manifesto on the Future of Seed." In Manifestos on the Future of Food and Seed, 76-102. Cambridge, MA: South End Press, 2007.

--. Stolen Harvest: The Hijacking of the Global Food Supply. Cambridge, MA: South End Press, 2000.

Slow Food International. GMOs. http://www.slowfood.com/international/22/gmos.

--. Food Sovereignty. http://www.slowfood.com/international/16/food-sovereignty.

--. Our Philosophy. http://www.slowfood.com/international/2/our- philosophy.

Slow Food USA. Ark of Taste. http://www.slowfoodusa.org/index.php/programs/details/ark_of_taste/.

--.Campaigns. http://www.slowfoodusa.org/index.php/programs/details/campaigns/.

Sustainable Agriculture Research and Education Program. ASI Research. http://www.sarep.ucdavis .edu/research.

--. Defining Sustainable Community Food Systems.

http://www.sarep.ucdavis.edu/sarep/sfs/def.
Tolford Wilson, Mary. "Peaceful Integration: The Owner's Adoption of His Slaves' Food." Journal of Negro History 49, no. 2 (1964): 116-127.
United States Department of Agriculture. Farmers Markets and Local Food Marketing. http://www.ams.usda.gov/AMSv1.0/foodhubs

Chapter 9: Trends in Local Food Production: The Growing Importance and Impacts of Community Supported Agriculture in the United States

Christopher D. Merrett, Frederick Iutzi, Erin Orwig, Sarah Shike, and Heather McIlvaine-Newsad

Introduction

Over the past three decades, American consumers have witnessed the growing popularity of locally-grown foods. The increasing interest in local foods is driven by several factors, including worries about food safety and environmental damage related to industrial agriculture, the pursuit for better quality foods that have become associated with local producers, and concern that the industrial model of agriculture has harmed rural communities through farm consolidation and population outmigration. In order to better understand the current and future role of the local foods phenomenon, this chapter examines a particular model of local food production called Community Supported Agriculture (CSA). CSA is a consumer cooperative where multiple households purchase shares in a local farm at the beginning of each season. In exchange for the initial investment, the farmer delivers a basket of seasonally appropriate vegetables, fruits, and other produce to each investor for the duration of the growing season. The CSA model is an important part of the expanding local food movement in North America, which emphasizes "civic agriculture" rather than an "industrial agriculture" paradigm (Lyson 2004). In order to learn more about this phenomenon, this chapter reports on two national surveys done in 2002 and 2009 on CSA farmers. The surveys show that CSA operators are motivated by concerns for fresh food, environmental sustainability, and community development. This chapter builds on previous research on local foods initiatives to explore the potential for local food systems to enhance community economic development efforts in rural America. The first section explores the rise of industrial agriculture. The second section argues that the local food movement is a form of "civic" agriculture that evolved in response to the industrial model of agriculture. Section three summarizes the two national surveys of CSA operators, focusing especially on the viability of CSAs to expand production. The final section examines the implications of CSAs for the way that we eat.

Industrial Agriculture

By many measures, the United States has the most productive agricultural system in the world. It is the number one producer of corn, and a top five producer of many other crops including wheat and soybeans. American farmers have more productive capacity than other country (Beddow, Pardey and Alston 2009). One positive and important consequence of this productivity is that Americans have some of the lowest food costs in the World. In fact, Americans spend less on food as a proportion of total household expenditures than any other country in the world (Meade 2012). According to the United States Department of Agriculture (USDA), Americans spend about 6.2 percent of their income on food. By comparison, the British spend 9.4 percent, Canadians spend 9.7 percent, and the Japanese spend 14.7 percent. Developing countries spend a much higher proportion of their household income on foods. In Latin America, Argentines spend 21.2 percent of their household income on food while Peruvians spend 36.6 percent. In Africa, Nigerians spend 39.6 percent of their household income on food.

Many reasons explain why American farmers produce abundant food at low cost. Two reasons stand out: mechanization of agriculture and farm subsidies. The application of mechanization and technology to agriculture has led to lower food costs, but has raised concerns about the impact of industrial agriculture on rural communities, the environment, food safety, and food quality. Cochrane (1993) explains how farmers who invest in new technology can get trapped on the so-called "technological treadmill." He argues that investment in new, costly technology does increase on-farm productivity, but it leads to farm consolidation, larger farms, fewer farmers, and rural depopulation. It creates this "treadmill" because farmers who invest in new technology early on increase the amount of food in the marketplace and secure a short term boost in profits (Levins and Cochrane 1996). However, the increased supply of food drives down commodity prices, cutting into the profit margins, especially as other lagging farmers invest in the new technology. Also, because the farmer usually takes on more debt to purchase the new technology, there is pressure to amortize the costs of the new technology over a larger farm size, which further accelerates farm consolidation. Additionally, the growing farm size and farm productivity further places downward pressure on commodity prices, which imposes more pressure on farmers to invest in new technology to boost profits (Cochrane 1993). Farmers who want to stay in business have to keep running on the technological treadmill. As the metaphor implies, the farmers are merely remaining in place despite big investments in technology. The long term outcome is cheaper foods, growing inequality between rich and poor farmers, and rural depopulation.

These new technologies, including the mechanization of agriculture, genetically modified organisms (GMOs), precision agriculture using GIS / GPS mapping technologies and variable rate sprayers, pesticides, and herbicides, significantly boosted agricultural productivity, and have been viewed by many observers of American agriculture as a success (Lanagcuster 2010). In fact, these new technologies boosted farm productivity growth above population growth rates, undermining Malthusian doomsayers who predicted massive famines would occur when population growth outpaced the growth of the food supply.

Despite the obvious successes of industrial agriculture, such as low-cost, abundant food, the intensive use of technology has raised significant concerns, many of which have spurred the rise of the local food movement. First, critics of industrial agriculture argue that large scale agriculture harms the environment. Chemicals such as nitrogen, applied to maintain soil productivity, run off fields into Midwestern streams and rivers, resulting in a "Dead Zone" at the mouth of the Mississippi River in the Gulf of Mexico (Scheer and Moss 2012). Other toxic farm chemicals seep into the water table and ecosystems contaminating ground water used for human consumption as well as wetlands inhabited by a range of plants and animals. Intensive cropping also accelerates soil loss and a reduction in biodiversity in regions where agriculture occurs.

Second, commentators question the impact of technology on food safety. Chemicals found in pesticides and herbicides remain on vegetables despite being washed or otherwise processed. Consumer advocates argue that GMOs represent a risk at the genetic level, with the potential to cause cancer and other diseases (Stier 2010; Worstell 2013). The quest for productivity (i.e. more output from fewer inputs) caused Mad Cow disease (i.e. Bovine Spongiform Encephalopathy, which occurred when herbivore beef cattle were fed meat and bone meal) and a range of harmful food-borne illnesses such as *listeria*, E. coli, and salmonella, which are linked to poor food handling, processing and storage techniques (Haiken 2011).

Third, a range of consumers, scientists, and chefs argue that food produced from industrial agriculture travels a long distance from the farm to the consumer. Food shipped over a long distance is not fresh, and in many instances is bred to endure rough handling rather than flavor and texture. In other words, the shift to industrial agriculture and longer supply chains has compromised food quality (Renting, Marsden and Banks 2003).

Finally, scholars of rural development suggest that farmers and communities are increasingly marginalized due to the global market pressures forcing farm consolidation (Berry 1999). Consequently, there are fewer, but larger farms. This has a ripple effect across the rural landscape. Farmers are part of the rural middle class. With fewer middle class members, there are fewer people attending rural churches, sending their children to local schools, consumers purchasing goods at local stores, and patients patronizing rural health care facilities. Technology-influenced farm consolidation causes demographic and economic stagnation in many parts of the Midwest and Great Plains (Johnson 2012). An estimated 56.3 percent of rural counties lost population between 2000 and 2008 (USDA 2011). Exacerbating these trends has been a farm subsidy system that favors large-scale commodity production such as corn, soybeans, wheat, dairy, and livestock over the production of fruits and vegetables (Rampell 2010). The bottom line is that industrial agriculture has decreased community size and vitality for many rural places.

The Rise of Civic Agriculture

There is a long tradition of direct farmer-to-consumer sales in the United States through farmer roadside stands and farmers markets. For example, Boston held its first farmers market in 1634 (Pyle 1971). Cincinnati started its farmers market in 1801. Dallas has had a farmers market since at least 1860. However, there has been a dramatic increase in local food systems over the past three decades. The USDA (2012a) estimates that local farmers markets grew from 1,755 in 1994 to 7,864 by 2012. Community Supported Agriculture, the subject of this chapter, also experienced dramatic growth over the past three decades. The first two CSAs are thought to have started in Massachusetts and New Hampshire in 1986. Current estimates put the total number of CSA farms at something greater than 6,000, although this number is subject to some debate (McFadden 2012; USDA 2012b). The emergence and increase in the 1980s of farmers markets and CSAs, exemplars of local foods systems, occurred in large part to address the perceived damage caused by industrial agriculture to our environment, bodies and palate.

CSAs are subscription-based buyers' cooperatives. Prior to the start of each growing season, subscribers purchase shares in a CSA. The farmer uses the money from subscribers to purchase the materials for the growing season including seeds, equipment and other inputs. In exchange for purchasing a share, the subscribers receive a seasonally-appropriate basket of vegetables, fruits, herbs and other produce for each week of the growing season. One outcome of the CSA model is that it helps farmers and consumers share the risk, or bounty, of the growing season. It provides a guaranteed income to the farmer regardless of the growing conditions in any given season. The food is fresh, grown locally and usually with few or no chemical inputs. Additionally, the CSA model allows consumers to get to know their farmer, and where their food comes from. CSA farms prompt consumers to meet each other too. Hence, it creates a sense of community around sustainable food production and consumption.

CSAs have the potential to address many of the problems caused by industrial agriculture. They require few if any chemical inputs, the food produced is fresh, and it is grown locally so that new social relationship can emerge around the production and consumption of food. This has prompted researchers to describe CSAs and other local food production and marketing systems as "civic agriculture" in contradistinction to "industrial agriculture" (Lyson 2000; DeLind 2002; Lyson 2004). Civic agriculture is akin to the "slow food movement" which says that the focus on agricultural productivity and, "...the 'Fast Life' has changed our way of being and threatens our environment and our landscapes" (Petrini 2004). As an *avant-garde* "riposte" to industrial agriculture, Carlo Petrini, founder of the Slow Food Movement, declares that farmers and consumers need to band together in "food communities" to create "places of gastronomic pleasure" (Petrini 1989; Pollan 2010). In doing so, people can ensure crop diversity, sustainable agriculture, and rural development.

Extending the work of Robert Putnam (1995) into the agricultural realm, Lyson (2000) recognizes in the Slow Food Movement the effort to build social capital through local food production. Local food systems promote civic engagement and draw our attention to our local environment and the human relationships that constitute our social network. Hence, Lyson (2000, 2004) and others rebrand local food systems as a form of "civic agriculture" which promote local communities as well as providing fresh food produced locally. It is an effort to diversify farms, as well as the rural economic landscape.

The rise of CSAs and other local food systems has prompted some farmers and policy makers to suggest that local food systems could even replace the industrial food system. Several states including Alaska, Maine, and Illinois have passed "Local Food, Farms and Jobs" legislation to get more locally produced food into schools and other institutional settings. Despite this support, there are still many obstacles to scaling up local food productivity (Saunders 2011). Large scale institutions such as schools, hospitals, and prisons, as well as commercial enterprises such as grocery stores need food suppliers to reliably provide food of sufficient quality and quantity for a set period of time. Most CSA operators lack the ability to meet these standards. It is not even clear that most local farmers even want to scale up their production. Scaling up production would entail greater complexity in production, food safety oversight, and possibly abandoning their existing community subscribers.

Surveys of CSA

In order to explore the motivations and operation of CSAs, two national surveys were conducted by the Illinois Institute for Rural Affairs (IIRA) in 2002 and again in 2009. The surveys focused on several themes including the basic operations of the CSAs, strategies for finding new subscribers and retaining old ones, and marketing strategies. These surveys shed some light on scalability and productivity issues to see if the CSA model can be part of a strategy to replace the old industrial paradigm.

Both surveys used paper surveys. An initial survey was mailed out to each address followed by a reminder postcard after two weeks. If no survey was received, a second survey was mailed to the CSA operator. The mailing list for conducting the surveys was generated from several sources including the Robyn Van En Directory (Van En 2002), internet searches, and local knowledge. In 2002, 746 viable respondents were identified and sent surveys. We received 373 surveys for a 50 percent response rate. In 2009, 633 surveys were mailed out. Responses were received from 190 for a 30 percent response rate

The first set of questions focused on the demographics and basic operations of the CSA (Table 1). Several findings stand out. First, CSA operators are on average over a decade younger than traditional farmers. The average age for a traditional farmer is about 58 years (USDA 2011). Second, women are much more likely to manage a CSA operation than a traditional farm. In the 2009 survey, 52 percent of CSA operators were women. Women only operate 14 percent of traditional farms (USDA 2007a). Wells and Gladwell (2000) suggest that the gender difference between CSAs and traditional farming is no accident. Traditional agriculture is characterized by patriarchy, competition, and individualism. CSAs offer an alternative form of agriculture that is more welcoming to women, with an emphasis on "care-practices" including "building relationships with other growers, shareholder-members, and the land" (Wells and Gladwell (2000, 107).

There is also a difference between CSA operators and traditional farmers in terms of educational attainment. Almost 80 percent of CSA operators had at least an undergraduate degree, with some having graduate degrees. In contrast, traditional farmers had lower levels of educational attainment. According to the USDA (2007b), there is a broad range of educational attainment for traditional farmers that is correlated to farm size and sales. At the lower end, about 25 percent of small family farm operators have an undergraduate degree. About 44 percent of large-scale farmers have an undergraduate degree (USDA 2007b).

Finally, CSA operations generate about 25 percent of the total household income for CSA households. In traditional farm households, farm income only generates about nine percent of farm household income (USDA 2011). The broader implications of these findings is that CSA operators are different than traditional farmers in terms of their average age, gender composition, educational attainment, and reliance on farm operations for household income.

CSA Attributes	**Mean 2002**	**Mean 2009**
Average Age of a CSA Operator / Farmer	44.6	46.1
Gender of CSA operators (% male / female)*	49 /51	48 /52
Educational Attainment (% with undergrad / graduate degree)	79.6	79.8
Were you raised on a farm (%)?	22.1	25.8
Avg. acreage of CSA in first Year	3.6	*
Avg. acreage of CSA in current year	8.6	*
Avg. # of subscribers in first season (Mean / Median)	26.9 / 20.0	*
Avg. # of subscribers in current season (Mean / Median)	91.7 / 50.0	*
Avg. price of subscription in first season ($)	318.75	*
Avg. price of subscription in current season ($)	399.72	*
Percent of total family income from CSA	25.1%	*

Table 1. Summary of CSA Operations
* Question not asked in comparable way in 2009.
Source: IIRA (2002, 2009).

A second set of questions focus on the factors motivating the CSA farmers to start their operations (Table 2). These questions asked respondents to rank the factors motivating them to start a CSA on a scale of 1 to 5, with 1 being not important and 5 being very important. A chi-square test was used to determine if the reported results for each question were non-random. Several results stand out.

Motivating Factors	**Mean 2002**	**Mean 2009**
Providing wholesome food to community	4.79**	4.70**

Promoting Sustainable agriculture	4.65**	4.45**
Land stewardship / Environmental concerns	4.53**	4.32**
Life style	4.32**	4.25**
Generating a sense of community	4.25**	4.34**
Profit	3.40**	3.47**
Keeping farmland within the family	3.11**	2.95**

Table 2. Factors Prompting Farmers to Start a CSA
** Results are significantly non-random at $p < .01$.
Respondents rated the factors on a scale of 1 to 5, where 1 = not important and 5 = very important. A chi-square test was used to determine if responses were significantly non-random.
Source: IIRA (2002, 2009).

First, the rank order of factors in 2002 was largely the same as in 2009. The importance of land stewardship increased one position in 2009, but overall, these results show that a fairly consistent set of values and motivations are inspiring farmers to operate CSAs. And these values, align very closely with the ideals promoted by the slow food movement and the values underpinning Lyson's (2000, 2004) conception of civic agriculture. There is a very strong emphasis placed on providing wholesome food, sustainability, and land stewardship. In 2009, the third highest ranked factor was "generating a sense of community." Hence, CSA operators understand that they are not merely producers of food. They also see themselves, to varying degrees, as builders of community. One local CSA operator has even described his CSA as a "third space," a venue beyond work and home where community members can congregate to share food and friendship (Curtis 2012; Peterson 2009). Note that in both 2002 and 2009, CSA operators only ranked profit as the sixth most important motivating factor. This finding could imply several things. First, it represents a different philosophical mindset than found in traditional farmers where profits rank much higher as a motivator. CSA farmers are not trapped on the technological treadmill. This seems like a laudable attribute. On the other hand, if advocates of local foods want to scale up local food production to increase availability of local foods for schools and the general public, CSA operators and other local food producers will have to pay more attention to efficiencies and profitability.

Finally, the survey asked if the CSA was a strategy to keep farmland in the family. The reason for asking this question is that many traditional farmers have been searching for additional profit centers on their farms. In order to generate more on-farm income farmers have been offering agritourism opportunities such as fee hunting (letting hunters on one's land for a fee), corn mazes, hiking, mountain biking, hay rides, vegetable stands and other strategies to bring consumers onto the farmland (Wicks and Merrett 2003). Some traditional farms have used CSAs as an additional income source. For example, Angelic Organics, currently the largest CSA in the United States was started in part as a strategy to save the family homestead. The story of Angelic Organics and its operator John Peterson was documented in an award-winning film called the *Real Dirt on Farmer John* (Peterson and Tiegart 2005).

Like any small business, CSAs take effort to operate and sustain. CSA operators face challenges such managing risk due to inclement weather, marketing to recruit subscribers, and working to maintain existing subscribers (Goland 2002). To better understand the challenges confronted by CSA operators, the survey asked farmers to rank a series of obstacles using a Likert Scale, where 1 equals not important and 5 equals very important (Table 3).

To begin, the obstacles were assigned surprisingly low scores given that previous research suggests that some CSAs experience as much as 50 percent subscriber turnover in some years (Kloppenberg et. al. 2000). Retaining subscribers was ranked as the most important obstacle in 2002, but its score dropped by 2009. While we cannot make direct linkages between 2002 and 2009 data because different CSA operators are responding, we can speculate that CSA operators might be learning how to respond better to subscriber concerns. Anecdotal evidence suggest that CSA operators are increasingly likely to have websites, newsletters, and end of year surveys that ask subscribers what worked and what didn't during the previous year. Greater attention to subscriber satisfaction through better communication might be having a positive impact on reducing subscriber turnover.

Obstacles	**Mean 2002**	**Mean 2009**
Retaining subscribers	3.32**	2.65**
Low profits	3.31**	2.77
Marketing	3.02*	2.31**
Variety of products offered	2.94	2.69*
Transportation issues (e.g. delivery)	2.91	2.69**
Competition with retail outlets (e.g. grocery stores)	2.43**	2.11**
Persuading subscribers to work on farm	2.12**	1.89**
Competition with other CSAs	1.85**	2.08**

Table 3. Ranking Obstacles to CSA Success
** Results are significantly non-random at $p < .01$.
* Results are significantly non-random at $p < .05$.
Respondents rated the factors on a scale of 1 to 5, where 1 = not important and 5 = very important. A chi-square test was used to determine if responses were significantly non-random.
Source: IIRA (2002, 2009).

Other scores include the low to moderate concerns for low profits and marketing. Given that profits were a lowly regarded motivating factor (Table 2), one would expect to see low profits as a minor obstacle. What was a surprise was the low and decreasing concern for marketing. The fact that marketing as a concern decreased between 2002 and 2009 might be attributable to the myriad newspaper articles, research reports, and movies covering the rise of CSAs over the past decade (Goland 2002; Peterson and Siegel 2005; Saulny 2008).

Running a CSA is hard work. In order to share the burden of operating the CSA, many CSAs request or even require that subscribers do some of the work to pick, clean, and sort the produce that CSA members take home. Survey results show that getting subscribers to work on the CSA is not a big problem. Results from 2002 and 2009 show, at least indirectly, the community building elements of CSAs as subscribers willingly do the work when asked.

The final line of inquiry posed above explores the impact of competition from other local food producers as well as grocery stores (Table 3). In the beginning of the local food movement, local food advocates touted farmers markets and CSAs as a way to help farmers avoid increasingly fierce global competition. CSAs were also viewed as an alternative to grocery stores and hence not likely to be in direct competition with grocery stores for consumers who want locally-grown, organic produce. Of course, once grocery stores saw the size of the untapped market, many grocery chains, including Wal-Mart, did increase the amount of locally-grown food for sale in their stores (Clifford 2010). The Midwestern grocery store chain, Hy-Vee, actually has a map on its website showing the location of the farmers who supply locally-grown foods to Hy-Vee stores (Hy-Vee 2013). The seemingly paradoxical implication here is that multi-state and even multi-national retail chains find descriptors like "sustainable," "organic," and "locally-grown" to have powerful marketing value.

Despite the rapid inclusion of organic and locally-grown foods into their inventories, large grocery stores chains are not perceived to be a large source of competition for CSAs in this survey (Table 3). CSAs in both surveys also do not view other CSAs as a competitive threat. Some studies have suggested that CSAs monitor nearby CSAs to make sure they are offering competitive prices (Ernst and Woods 2009). However, the overall sense is that CSAs are not engaged in cutthroat competition for clients or market share, unlike the competitive landscape for grocery stores.

In the 2002 survey of CSA operators, retaining subscribers was the highest ranked obstacle to CSA success. The 2009 respondents ranked subscriber turnover lower, but it still ranked as a moderately significant issue. Hence, the issue warrants closer scrutiny. Previous studies have examined the causes of subscriber turnover by surveying CSA members. Some CSAs have experienced turnover rates exceeding 50 percent in any given year (Goland 2002). Other studies report turnover rates ranging from 20 to 37 percent (Bruch and Ernst 2010, 7). High turnover rates are especially true during the early stages of a CSA's development as the farmer struggles to establish the farm, hone his or her skills, identify the right product mix, and so forth. A survey of 24 subscribers in one fledgling CSA in central Ohio asked participants at the end of the season whether their expectations had been met (Goland 2002, 20). Six (25%) reported having no expectations, two (8%) reported that their expectations had been met, and three (13%) felt that the CSA exceeded their expectations. Unfortunately, 13 (54%) reported that the CSA failed to meet their expectations (Goland 2002, 20). Reasons for subscriber dissatisfaction included inconvenience of having to prepare fresh foods, and having to eat unfamiliar foods. Of course, this goes to the very point of the Slow Food Movement. So many people eat fast food and pre-packaged grocery foods because of their convenience. Eating foods from CSAs can require subscribers to learn new recipes and menus in order to effectively consume all of the foods in the basket. Complicating the situation is the fact that the foods delivered through the CSA change during the course of the growing season.

In order to get an alternative view of the situation, the 2002 and 2009 surveys asked CSA operators why they felt subscriber turnover occurred. Results show that CSA operators understand the challenge of preparing fresh food (Table 4). CSA operators report that from their perspective, the lack of time for food preparation is a big factor, both in 2002 and 2009. Having to prepare unfamiliar varieties of food is also a challenge. Many CSA operators try to reduce these problems by providing recipes and newsletters to educate subscribers. An ongoing logistical problem recognized by CSA operators, is the nuisance of having to travel to pick up the basket of food each week. This reconfirms the notion that many CSA subscribers are naïve about what it takes to eat a healthy diet. It is not simply a matter of purchasing fresh food. It also requires that consumers rethink household schedules and tasks to cook healthy foods over the long term. The inability to make these larger changes contributes to subscriber turnover.

Reason for Not Renewing Subscription	Mean 2002	Mean 2009
Subscriber does not have enough time to prepare food	3.52**	3.15**
Subscriber moved away	3.37**	3.59**
Inconvenience	3.14**	2.68
Subscriber thought there was too much food	3.08	3.22
Too far for pick up or delivery	2.63**	2.48*
Unfamiliar varieties of food	2.62**	2.56**
Don't know	2.62*	2.67
Not enough product choice	1.92**	1.76**
Does not feel price is a good value	1.83**	1.76**
Subscriber thought there was not enough food	1.65**	1.62**

Table 4. Perceived Reasons Explaining Why People do not Renew CSA Subscriptions

** Results are significantly non-random at $p < .01$.

* Results are significantly non-random at $p < .05$.

Respondents rated the factors on a scale of 1 to 5, where 1 = not important and 5 = very important. A chi-square test was used to determine if responses were significantly non-random.

Source: IIRA (2002, 2009).

The fact that subscriber turnover is an inherent, ongoing part of operating a CSA means that CSAs must regularly recruit new subscribers. Having a marketing plan is a key part of maintaining a viable subscriber base (Bruch and Ernst 2010). The 2002 and 2009 surveys asked CSA operators to identify their various marketing strategies (Table 5).

Marketing Strategy	Percent Using 2002	Percent Using 2009
Word of Mouth	88.8	97.8
Internet Webpage	44.4	79.6
Farmers' Market	38.2	35.8

Involved in Community Events (e.g. County fair)	35.6	40.9
Open House—Inviting Community to Visit	35.3	31.4
Newspaper Advertisements	27.5	21.9
Roadside Stands	9.1	5.8
Radio	7.2	8.8

Table 5. Marketing Strategies used by CSA Operators
Source: IIRA (2002, 2009).

In both the 2002 and 2009 surveys, CSA farmers reported using a combination of marketing strategies. These strategies include "word of mouth," face-to-face strategies including farmers markets and roadside stands, traditional media advertising through newspapers and radio, as well as digital outlets such as having a webpage. The survey generated some noteworthy results. First, the results show a dramatic increase in using the internet to market the CSA. This is not surprising given the spread of broadband technologies through society. A second noteworthy result is that despite the growing reliance on the internet and marketing at a distance, CSAs continue to place a high priority on marketing locally through personal contacts and community engagement.

Discussion and Conclusions

The results of the two surveys show the ongoing commitment CSA operators have to local food and the role played by CSAs in community development. This support for local food mirrors the growing support consumers in general have for fresh, safe food produced locally. Building on this momentum is a growing push for expanded local food production through CSAs and other local food production and marketing enterprises (Neal 2013; Wallstin 2013).

Before this expansion can occur, however, CSA operators, lawmakers, and community stakeholders will have to reduce several barriers to expanded local food production. First, there are several internal issues to resolve within the CSA as a business model. The 2002 and 2009 surveys revealed that subscriber turnover is a significant and ongoing issue. Many CSA operators accept subscriber turnover as an inevitable part of doing business, but replacing departing subscribers takes energy, time and money that could be better used elsewhere in CSA operations, including expanding productivity.

The lack of business planning is another constraint on current CSA success and plans for future expansion. Results from the 2009 survey reveal that only 50 percent of the CSAs had a business plan (IIRA 2009). It will be difficult for CSAs to expand operations without having a better handle on their operations including cash flow, profitability, and return on initial investments in land, equipment and so forth. Part of having a good business plan is having a marketing plan. And as previously noted, CSAs have to market year round to replenish their roster of subscribers. However, the fact that a large number of CSAs don't have a business plan and don't have a well-designed marketing strategy raises questions about the ability of local producers to increase productivity.

A final obstacle to expanded CSA production is the willingness and motivation of CSA operators to move beyond their niche operations. The survey results show that the profit motive ranks lower than lifestyle and concern for community as factors motivating the farmers to run their CSAs. Given slim or non-existent profit margins caused by the lack of attention paid to the bottom line, it will be difficult for CSAs to contribute to expanded demand for local foods. Expanding CSA operations would also likely require increased labor. An underappreciated aspect of running a CSA is that it can be hard physical labor. The hard work can even result in overuse and repetitive motion injuries (Boeke 2010). Two CSA operators commented in the 2009 survey that they were "currently maxed out" and could not conceive of expanding their operations (IIRA 2009). In other words, there are human limits on how much a CSA can produce.

There are also external barriers to expanded local food production from CSAs. Some CSA operators do think entrepreneurially. For example, some CSA operators enhance their own product line by including locally produced processed foods such as salsa, jams, breads, and even meats. However, there are federal (USDA) food safety regulations in place to govern the sale of processed foods. Large-scale food processors can afford the equipment and inspectors required to comply with these standards. Small-scale producers such as CSAs usually find these added expenses cost-prohibitive, and barriers to expanded and diversified production.

There are also supply chain constraints to expanded production. Some CSAs might consider selling some of their produce to grocery stores, restaurants, or government institutions such as schools or hospitals. As a condition of supplying food to these large scale enterprises, CSAs have to meet minimum standards for food quality as well as quantity delivered over a prescribed time period. Most retailers or institutional purchasers prefer to lock in long term contracts rather than deal with multiple small-scale suppliers. Few CSAs could meet these requirements by themselves without somehow increasing their productivity.

The challenge of expanding CSA productivity could potentially be alleviated by having greater access to low-interest loans which could allow the CSA operator to purchase more land, hire employees, or invest in new equipment and facilities (e.g. hothouses to extend the growing season in cooler climates, or irrigation systems to expand production in drier, warmer climates). However, many CSAs would not qualify for loans from traditional lenders such as banks. They have to rely on federal and state government programs (e.g. USDA) which have limited funding.

A final external factor to consider is that CSA productivity is affected by its geography. Regional variations in climate, soil quality, water availability and market accessibility among other factors will affect the ability of a CSA to expand production if its operator wanted to do so. Just consider the potential of a CSA operating in Minnesota versus one operating in California. Seasonality, climate, and accessibility to large urban markets are very different in these locations.

Despite these barriers, it is clear that progress is happening. Research, technical assistance outreach and policy changes are occurring at the federal, state, and local levels to increase the productivity of CSAs and expand the access American consumers have to locally produced food. At the federal level, the USDA has implemented several new initiatives to promote CSAs and other local food producers. For the first time in 2007, the USDA began to count CSAs in the Census of Agriculture (USDA 2007c). Tracking CSAs will help provide a benchmark for future policy decisions related to local food production. In 2009, the USDA started the "know your farmer, know your food" initiative that includes online maps, directories, and technical assistance to promote CSA, farmers markets, and other local foods initiatives. The USDA has also expanded the funding available to local foods initiatives including the Farmers Market Promotion Program, Farm to Schools Grants, and the Community Food Projects Competitive Grant Program, among many other grant and lending programs (Fitzgerald, Evans, and Daniel 2010).

Many states are also making it easier for local food producers, including CSAs, to expand their production and profitability. For example, in 2010, Illinois passed the Local Food, Farms, and Jobs Act. It aims to increase the demand for CSA output by creating a market for locally grown foods in the public school system. Another concrete goal of this legislation is to have state agencies that serve food (e.g. prisons) secure at least 20 percent of food from local growers by the year 2020.

In January 2012, Illinois promulgated the Local Food Entrepreneur and Cottage Food Operation Act which makes it easier for local producers in Illinois to process and sell limited amounts of processed food. This could spur entrepreneurship among local farmers who can reach up the value-chain to produce jams, jellies, pesto, salsa and other products in order to add value to the fruits and vegetables that they grow.

State-supported institutions of higher education are also promoting local foods. For example, John Woods Community College in Quincy, Illinois, offers an associate degree in sustainable agriculture. The University of New Hampshire and the University of California—Davis both offer undergraduate degrees in sustainable agriculture and foods systems.

Local foods advocates are also working diligently at the local level. Local communities are building food business incubators and community kitchens that comply with health and safety standards to encourage local entrepreneurs to process foods grown locally (Pulliam 2011). In order to share the burden of marketing and expanding the variety of products sold, CSAs are forming cooperatives. For example, the Good Earth Food Alliance located in West Central Illinois is comprised of seven CSAs. Each CSA specializes in particular produce (including pesto, eggs, and poultry), thereby expanding the total range of products available to CSA subscribers (Good Earth Food Alliance 2013). Another strategy used by local food producers to meet increased demand is to create food hubs. Where one CSA or local food producer might not be able to meet the supply needs for an institutional purchaser, several local food producers or CSAs working together might be able to meet the demand by combining their productive capacity. This strategy allows farmers to scale up production while remaining small-scale producers.

In addition to activities taking place independently at the federal, state, and local levels, food interest groups are also emerging that transcend scale. For example, the "buy fresh, buy local" initiative has a national organizational structure, but has regional and local chapters to reflect the unique food regions across the United States (Food Routes Network 2013).

Despite considerable recent success for CSAs and other local producers, an important question remains to be answered: is scalability possible? CSAs and other local food producers have a serious focus on the environment, food safety, and community, but a less serious focus on economics, profitability, and supply chain issues. Can CSAs and other forms of local food production offer us the sustainability benefits of "civic agriculture" while providing the food abundance of "industrial agriculture"? A tantalizing, potential model might be found in Denmark. There, the Aarstiderne CSA delivers food each week to 45,000 subscribers in Denmark and another 5,000 subscribers in Sweden (Black 2009). Thomas Harttung, the CSA operator sources about 70 percent of the food from within Denmark. About 35 percent comes directly from Harttung's farm. The Aarstiderne CSA was started in 1999. By 2009 it was generating about $45 million in annual revenues. In order to scale up production to achieve this success, Harttung (quoted in Black 2009) said"...we had to give up some of the high-church ethics of a CSA and adopt a more ecumenical approach... to really change things, we cannot preach to a very small congregation. We must deliver a message about organic food to a broader audience." The immediate implication is that a rigid adherence to ecological values would have prevented him from achieving the economies of scale required to increase production.

The broader implication is that he had to strike a balance between idealism and practicality. Some people might accuse Harttung of compromising his values. *Time Magazine*, on the other hand, identified Harttung as one its 2009 "heroes of the environment" (Abend 2009). Will American CSA operators feel as though they are compromising their values in their quest to scale up production? The survey results in this study show that American CSA operators prioritize environmental and community ideals over the economic bottom line. Resolving this tension will require further work at the federal, state, and local level by farmers, consumers, elected officials, government agencies, food retailers, and other stakeholders. As Farmer John says in the movie *The Real Dirt on Farmer John*, "It ain't easy being green" (Peterson and Siegel 2005).

References

Abend, Lisa. 2009. Heroes of the environment 2009: Thomas Harttung. *Time Magazine*. September 22, pg. 84. Available online at: http://www.time.com/time/specials/packages/article/0,28804,1924149_1924155_1924437,00.html. Accessed: January 25, 2013.

Beddow, Jason M., Philip G. Pardey, and Julian M. Alston. 2009. The shifting global patterns of agricultural productivity. *Choices Magazine*. 24 (4): 1-10.

Berry, Wendell. 1999 The death of rural communities. *The Ecologist*. 29(3):183-184.

Black, Jane. 2009. The CSA that isn't one. *Gourmet*. January 6. Available online at: http://www.gourmet.com/foodpolitics/2009/01/danish-csas. Accessed: January 15, 2013.

Boeke, Diana. 2010. The vegetables of labor. *The Clueless Farmer* (Blog). June 20. Available online at: http://cluelessfarmer.wordpress.com/2010/06/20/vegetables-of-labor/. Accessed: February 6, 2013.

Bruch, Megan and Matt Ernst. 2010. *A Farmer's Guide to Marketing through* Community *Supported Agriculture (CSAs)*. Center for Profitable Agriculture, University of Tennessee, Knoxville, TN. Available online at: https://utextension.tennessee.edu/publications/Documents/PB1797.pdf. Accessed: February 6, 2013.

Clifford, Stephanie. 2010. Wal-Mart to buy more local produce. *New York Times*. October 14. Available online at: http://www.nytimes.com/2010/10/15/business/15walmart.html?_r=0. Accessed: January 15, 2013.

Cochrane, Willard. 1993. The Development of American Agriculture: A Historical Analysis, Second Edition. Minneapolis, MN: University of Minnesota Press.

Curtis, John. 2012. Personal conversation with operator of Barefoot Gardens, CSA located in Macomb, Illinois, Saturday, August 11, 2012.

DeLind, Laura. 2002. Place, work, and civic agriculture: Common fields for cultivation. *Agriculture and Human Values*. 19: 217–224.

Ernst, Matt and Tim Woods. 2009. *Community Supported Agriculture (CSA)*. UK Cooperative Extension Service. Available online at: http://www.uky.edu/Ag/NewCrops/marketing/csa.pdf Accessed: January 20, 2013.

Fitzgerald, Kate, Lucy Evans, and Jessica Daniel. 2010. *The National Sustainable Agriculture Coalition's Guide to USDA Funding for Local and Regional Food Systems, First Edition*. Washington, DC: National Sustainable Agriculture Coalition. Available online at: http://sustainableagriculture.net/wp-content/uploads/2010/06/6.18-FINAL-Food-System-Funding-Guide2.pdf. Accessed: February 6, 2013.

Food Routes Network. 2013. Buy Fresh, Buy Local. Available online at: http://foodroutes.org/buy-fresh-buy-local/. Accessed: February 6, 2013.

Goland, Carol. 2002. Community supported agriculture, food consumption patterns, and member commitment. *Culture & Agriculture*. 24 (1): 14-22.

Good Earth Food Alliance. 2013. Grower / Producer Profiles. Available online at: http://www.goodearthfoodalliance.com/growerproducer-profiles. Accessed: January 28, 2013.

Haiken, Melanie. 2011. The 5 deadliest food-borne illnesses - and how to prevent them. *Forbes Magazine*. Available online at: http://www.forbes.com/sites/melaniehaiken/2011 /09/28/the-5-deadliest-food-borne-illnesses-and-how-to-prevent-them/. Accessed: January 31, 2013.

Hy-Vee. 2013. Home Grown Map. Available online at: http://www.hy-vee.com/resources/homegrown.aspx. Accessed: January 31, 2013.

Illinois Institute for Rural Affairs (IIRA). 2009. Survey of Community Supported Agriculture (CSA) Farms. Illinois Institute for Rural Affairs, Western Illinois University, Macomb, IL.

Illinois Institute for Rural Affairs (IIRA). 2002. Survey of Community Supported Agriculture (CSA) Farms. Illinois Institute for Rural Affairs, Western Illinois University, Macomb, IL.

Johnson, Kenneth. 2012. *Rural Demographic Change in the New Century: Slower Growth, Increased Diversity*. Carsey Institute Issue Briefing No. 44. Available online at: http://www.carseyinstitute.unh.edu/publications/IB-Johnson-Rural-Demographic-Trends.pdf. Accessed: February 1, 2013.

Kloppenburg, Jack, Jr., Sharon Lezberg, Kathy DeMaster, G.W. Stevenson, and John Hendrickson. 2000. Tasting food, tasting sustainability: Defining the attributes of an alternative food system with competent, ordinary people. *Human Organization*. 59 (2): 177-86.

Lanagcuster, Jim. 2010. Farming treadmill highly successful. *Southeast Farm Press*. September 3. Available online at: http://southeastfarmpress.com/management/farming-treadmill-highly-successful. Accessed: January 31, 2013.

Levins, Richard and Willard Cochrane. 1996. The treadmill revisited. *Land Economics*. 72 (4): 550-553.

Lyson, Thomas. 2004. *Civic Agriculture: Reconnecting Farm, Food, and Community*. Medford, MA: Tufts University Press.

Lyson, Thomas. 2000. Moving toward civic agriculture. *Choices Magazine*. 15 (3): 42–45.

Neal, Dale. 2013. USDA has loans for farmers growing for locals. *Asheville (NC) Citizen-Times*. January 20. Available online at: http://www.citizen-times.com/article/20130121/NEWS/301150044/USDA-offering-loans-farmers-who-grow-locals. Accessed: February 6, 2013.

McFadden, Steven. 2012. Unraveling the CSA Number Conundrum. *The Call of the Land: An Agrarian Primer for the 21st Century*. Available online at: http://thecalloftheland.wordpress.com/2012/01/09/unraveling-the-csa-number-conundrum/. Accessed: February 1, 2012.

Meade, Birgit. 2012. Percent of household final consumption expenditures spent on food, alcoholic beverages, and tobacco that were consumed at home, by selected countries, 2011. Economic Research Service. United States Department of Agriculture. Available online at: http://www.ers.usda.gov/data-products/food-expenditures.aspx#26654. Accessed: January 30, 2013.

Peterson, Christopher. 2009. Happy places: Third places. *Psychology Today*. December 1. Available online at: http://www.psychologytoday.com/blog/the-good-life/200912/happy-places-third-places. Accessed: 1/26/2013.

Peterson, John and Taggart Siegel. 2005. *The Real Dirt on Farmer John* (Film). San Francisco, CA: Collective Eye Productions.

Petrini, Carlo. 2004. *Slow Food: The Case for Taste*. Translated by William McCuaig. New York, NY: Columbia University Press.

Petrini, Carlo. 1989. Slow Food Manifesto. Available online at: http://www.slowfood.com/international/2/our-philosophy. Accessed: February 4, 2013.

Pollan, Michael. 2010. The food movement, rising. *The New York Review of Books*. May 20. Available online at: http://michaelpollan.com/articles-archive/the-food-movement-rising/. Accessed: January 28, 2013.

Pulliam, John. 2011. Open house to introduce Galesburg community kitchen at SBC. *Galesburg (IL) Register Mail*. January 27. Available online at: http://www.galesburg.com/news/ x2069925300/Open-house-to-introduce-Galesburg-community-kitchen-at-SBC. Accessed: January 28, 2013.

Putnam, Robert. 1995. Bowling alone: America's declining social capital. *Journal of Democracy*. 6 (1): 65-78.

Pyle, Jane. 1971. Farmers' markets in the United States: Functional anachronisms. *Geographical Review*. 61 (2): 167-197.

Rampell, Catherine. 2010. Why a Big Mac costs less than a salad. *New York Times*. February 4. Available online at: http://economix.blogs.nytimes.com/2010/03/09/why-a-big-mac-costs-less-than-a-salad/. Accessed: January 15, 2013.

Renting, Henk, Terry Marsden, and Jo Banks. 2003. Understanding alternative food networks: Exploring the role of short food supply chains in rural development. *Environment and Planning A*. 35: 393–411.

Saulny, Susan. 2008. Cutting out the middlemen, shoppers buy slices of farms. *New York Times*. July 10. Available online at: http://www.nytimes.com/2008/07/10/us/10farms.html? pagewanted=all. Accessed: February 2, 2013.

Saunders, Rhys. 2011. More local food in school lunches: Not a simple process. *State Journal-Register*. October 18. Available online at: http://www.sj-r.com/topstories/ x984142183/More-local-food-in-school-lunches-Not-a-simple-process. Accessed: January 15, 2013.

Scheer, Roddy and Doug Moss. 2012. What causes ocean "dead zones"? *Scientific American*. September 25. Available online at: http://www.scientificamerican.com/article.cfm? id=ocean-dead-zones. Accessed: February 19, 2013.

Stier, Ken. 2010. How Frankenfood prevailed. *Time Magazine*. June 28. Available online at: http://www.time.com/time/magazine/article/0,9171,1997448,00.html. Accessed: February 3. 2013.

United States Department of Agriculture. 2012a. Farmers Market Growth: 1994 to 2012. Available online at: http://www.ams.usda.gov/AMSv1.0/farmersmarkets. Accessed: February 2, 2013.

United States Department of Agriculture. 2012b. Community Supported Agriculture. Available online at:

http://www.nal.usda.gov/afsic/pubs/csa/csa.shtml. Accessed: January 31, 2013.

United States Department of Agriculture. 2011. Briefing on the Status of Rural America. Available online at: http://www.usda.gov/documents/Briefing_on_the_Status_of_Rural_America_Low_Res_Cover_update_map.pdf. Accessed: January 17, 2013.

United States Department of Agriculture. 2007a. 2007 Census of Agriculture: Women Farmers. Available online at: http://www.agcensus.usda.gov/Publications/2007/Online_Highlights/Fact_heets/Demographics/women.pdf. Accessed: January 20, 2013.

United States Department of Agriculture. 2007b. Structure and Finances of U.S. Farms: Family Farm Report, 2007 Edition / EIB-24. USDA Economic Research Service. Available online at: http://www.ers.usda.gov/media/201435/eib24c_1_.pdf. Accessed: February 5, 2013.

United States Department of Agriculture. 2007c. *2007 Census of Agriculture, Full Report, Volume 1*. Table 44, Selected Practices, page 606. Available online at: http://www.agcensus.usda.gov/Publications/2007/Full_Report/Volume_1,_Chapter_2_US_State_Level/st99_2_044_044.pdf. Accessed: January 31, 2013.

Van En, Robyn. 2002. Directory of CSAs. Available online at: http://www.wilson.edu/about-wilson-college/fulton/robyn-van-en-center/csa-search/index.aspx. Accessed: February 14, 2013.

Wallstin, Brian. 2013. Lawmakers Propose Farm-To-Table Program to Promote Local Agriculture. New Hampshire Public Radio. Available online at: http://stateimpact.npr.org/new-hampshire/2013/02/05/lawmakers-propose-farm-to-table-program-to-promote-local-agriculture/. Accessed: February 6, 2013.

Wells, Betty and Shelly Gladwell. 2000. Gender and resource management: Community supported agriculture as caring-practice. *Agriculture and Human Values*. 18: 107–119.

Wicks, Bruce and Christopher D. Merrett. 2003. Agritourism: An economic opportunity for Illinois. *Rural Research Report*. 14 (9): 1-8. Available online at: http://www.iira.org/pubs/publications/IIRA_RRR_577.pdf. Accessed: December 1, 2012.

Worstell, Tim. 2013. How those GM Frankenfoods are going to murder us all in our beds. *The Telegraph*. February 4. Available online at: http://blogs.telegraph.co.uk/finance/timworstall/100020189/how-those-gm-frankenfoods-are-going-to-murder-us-all-in-our-beds/. Accessed: February 4, 2013.

Chapter 10: Let the Flavors Marry: A Gender-Integrated Conceptualization of Nourishment for All People

Holly Stovall

Twenty-first century feminists, with their vision of justice for all peoples, have an important role in critiquing the food system and pointing towards more nourishing and just ways of eating that integrate what have been conceptually separated: masculinity and femininity, the lab and the kitchen, science and intimacy. The links between industrial foods and disease--such as diabetes, heart disease, kidney stones, and cancer—are growing stronger by various measures and are widely documented. Because choosing and providing nourishing food is associated with the traditionally stifling domestic sphere, some feminists have been hesitant to examine the theoretical connections between nutritional justice and feminism. At the same time, feminist theorists in the tradition of deconstruction have not specifically addressed nutrition science, though Evelyn Fox Keller has uncovered the masculine bias behind institutional science in general.

Institutional science has evolved under what Evelyn Fox Keller says is "the formative influence of a particular ideal of masculinity" (7). This ideal, says Keller, maintains a strict division between the constructions of masculinity and femininity, between the objective and subjective, power and love, and reason and intuition. Because nutrition science is a branch of institutional science, Keller's theory applies to nutrition science. Nutrition science privileges the masculine side of this dichotomy: objectivity and reason. By contrast, to value women's historical food and health knowledge, often assumed to be intuition, is understood as privileging the feminine side of the dichotomy. Keller argues that a more effective scientific process would dispense with the gendered polarities, and instead integrate the objective and subjective, reason and intuition into the way we do science (7-9), or in this case, the way we determine how to use food to sustain human beings and the planet. A feminist vision of nourishment will integrate old knowledge with new science. The effect is to mix and stir together what conceptually (and often in practice) has been held separate.

In addition to integrating the kitchen and the lab, a feminist discourse of nutrition science will conceptually unite the subject who consumes food for nourishment with the food itself. Keller argues that traditional science is rooted in a discourse that constructs a strict division between the "knower" (the mind, associated with masculinity) and the "knowable" (nature, associated with femininity) (1985, 79). The "knower" must remain above and separate from the "knowable." There must be no intermingling or intimacy—no mixing and stirring. With nutrition, however, the scientist as "knower" takes the "knowable" into her or his body, for a scientist can't survive without food. Moreover, the "knowable" and the "knower" literally become one, making it impossible to avoid intermingling in any concrete way. A conceptual separation, however, may be maintained. A feminist approach to nutrition science would conceptually acknowledge the inevitable union between the scientist and food. The problem, as Wendell Berry and others have identified, is that in the United States we have separated ourselves from and made ourselves superior to (conceptually and in practice), the source of our food. Berry identified this separation, decades ago, as an agricultural and spiritual crisis (1977)-- this author identifies the conceptual separation of food from those who consume it for nourishment as a nutritional crisis as well.

Conceptual integration must occur between various "parties" that have been held apart, often hierarchically. Notably, the kitchen has been separated from the lab. Neither Keller nor this author is suggesting a rejection of institutionalized science. The problem is that we have seen a large-scale rejection of women's tried and tested knowledge developed in the home. Historically and currently, "science" has rejected the expertise of midwives, breastfeeding mothers ("formula" is "scientific," but mother's milk is not), women's cooking, and the wisdom of "old wives' tales." That chicken soup is good for a cold is an "old wives' tale" that institutional science has vindicated, but why must we wait for "science" to validate women's old knowledge? We would not think of the reverse: considering science to be invalid until proven effective by "old wives." One advantage of food knowledge passed through "old wives' tales" is that it does no harm. Scientists, however, told people to eat margarine for decades until it was found to be much more dangerous than butter. Similarly, infant formula was promoted as a scientifically complete substitute for breast milk until medical scientists identified the many short and long-term health problems for mother and infant that often arise when formula is used. Regardless of the level of the scientific community's knowledge of formula versus breast milk, women's old wisdom about how to make infants thrive on breast milk was disregarded. We need a new, feminist-compatible, science that integrates old knowledge, passed down through relationships, with institutionalized science. The old knowledge will act as a check on new, untested science. It will say, slow down find out more about how the body handles margarine or formula before science, as an institution, endorses it.

Keller's theory is primarily focuses on the hidden ideology behind science—not on the gender of the scientist or theorist. A man, such as Berry, might conceptualize the science of soil and food in a way that integrates masculinity and femininity, the "knower" and the "knowable." A woman may conceptualize "nutrition" in a more institutional way that maintains a separation. This point is important because historically, there have been women, even feminists, who have taught and practiced institutionalized nutrition science, but the fact that a woman or feminist practices institutionalized nutrition science does not mean she conceptaully integrates femininity and masculinity. For example, in 1885, Ellen Richards, an instructor in "sanitary chemistry" and arguably a feminist, published the first edition of *Food Materials and their Adulterations*, but her scientific references are men practicing institutional science: namely Albert Leach's *Food Inspection and Analysis* and, in the 1906 edition, Andrew Winton's *The Microscopy of Vegetable Foods*. In addition, Richards's place of work, the Massachusetts Institute of Technology, was heavily male-dominated (Hayden, 157). Practicing nutrition science in this male setting is likely to reinforce masculine ideals of science more so than women practicing at the "hearth."

As a model of her theory, Keller documents how 20th-Century scientist Barbara McClintock was able to practice science in a way that integrated masculinity and femininity. McClintock won many awards, including a Nobel Prize for her discoveries about DNA, but because of the alternative scientific discourse she practiced, the scientific community failed to adequately recognize her work for many years (Keller, 1983, xviii). In her biography of McClintock, Keller emphasizes a scientific practice of relationships and intimacy, especially between the "scientist" and "nature." McClintock, says Keller, sought an "intimate and total knowledge" about "each and every plant" (1983, 105). It was this intimate approach that led both to her rejection by the scientific community and her groundbreaking discoveries about DNA.

It is important to envision the home cook, a daily "scientist," similarly: to what extent, historically, has the home cook sought intimate and total knowledge about her or his ingredients and the cooking process, as well as an intimate knowledge of how humans respond, over the short and long term, to that food? When over-processed, over-packaged, and over-marketed foods were not available, the home cook was forced to pass more time in the kitchen and therefore developed a much more intimate and thorough knowledge about the "nature" of food, cooking, and well-being. This element of daily, intimate, scientific practice in the home is one of the components missing from the discourse of institutionalized nutrition science. Unlike the scientist in the lab, the home cook as scientist has the opportunity to engage in an intimate relationship with the people who eat her/his food—she or he knows what causes a stomach ache or an altered mood. According to Keller's theory, if our understanding of human nutrition is to evolve, the theories and conclusions of daily cooks who have the opportunity to practice intimacy with cooking and with the people who eat the food must be integrated into the discourse and practice of science.

Keller accentuates McClintock's capacity for "a deep reverence for nature, a capacity for union with that which is to be known" which "reflect a different image of science from that of a purely rational enterprise" (1983, 201). Integrating Keller's concepts of reverence and union into nutrition science will entail an emphasis on feelings: how we feel about the food we cook and eat, about the people who grew and produced our food, and, importantly, about the soil our food came from. A pleasing taste, as well, might be recognized as a nutritionally important concept: one study shows that when people eat food that is pleasing to them, they are more likely to absorb the nutrients (Hallberg). Taste is subjective and tied to feelings and experiences, elements that institutionalized science has not routinely integrated.

A strong connection does, indeed, exist between feminism and the sense of confidence, satisfaction, accomplishment that comes from cooking and knowing the source of one's food. Stovall, Baker-Sperry, and Dallinger have documented that a high proportion of feminists are involved in one local foods movement, and that these feminists believe cooking and nutrition are important. These feminist local foods activists are part of a progressive community. Many of them teach Women's Studies or Women's History. Some have never married, while others are married with children and are the primary wage earners in their families. None of them, in this survey, fulfill the conservative role of the stay-at-home mom. Stovall's data provide an alternative conclusion to the one Tracy Deutsch makes about the local foods movement. Deutsch argues that the local foods movement "forwards a conservative set of assumptions about women's history and, by extension, women's present responsibilities" (168). A majority of those surveyed in the Stovall study reported that men contribute in a balanced way to childrearing, cooking, and cleaning. Many men in the survey cook and do not believe that nourishment is "women's responsibility."

Popularized Betty Friedan-based feminist principles that exhort women to leave the kitchen and delegate the important work of food knowledge to the "experts" may easily conflict with principles of eco-feminism and the theories of women's history, which emphasize women's agency and areas of expertise. Rather than calling only women back to the kitchen, feminists in the local foods movement call men there as well. Other feminists in the movement simply enjoy cooking and consider it their privilege and prerogative. It is too easy to relegate food choices to the realm of "light feminism," which stresses do-it-yourself activities and women's prerogative to "choose" to stay home. All women eat. Feminists, however, are called to eat in a way that's consistent with justice for all people and the planet.

Procuring their food mostly from supermarkets after World War II, Americans of the United States have unwittingly turned control of nourishment over to profit-driven industrialized food production and settled, often unknowingly, for cheaply and toxically generated foods. As eaters and gardeners, if not cooks, men as much as women bear responsibility. Before World War II, we ate minimally processed foods grown or raised with no chemicals or hormones. Wheat and tomatoes had not been hybridized to make them more easily harvested with machines. Nothing was genetically modified. Sodas, beverages, and treats had not yet been made cheap with high fructose corn syrup. Livestock wasn't caged or routinely fed corn, antibiotics, or hormones. Canned goods were not lines with estrogen-mimicking plastics. Scholars, farmers, and journalists have documented many more differences in the food supply. In short, it was impossible for pre-World War II Americans to eat the way we do today. Correspondingly, they did not suffer the epidemics of diet-caused disease.

Mildred Armstrong Kalish's recent memoir, *Little Heathens: Hard Times and High Spirits on an Iowa Farm During the Great Depression* provides many examples of consumers of food for nourishment engaging in an intimate relationship with their food. This memoir is important because it tells a story of a healthy way of eating that does not place the nutritional burden on women alone. Moreover, her story is compatible with eco-feminist thought, which advocates small, diverse, ecologically-responsible farms and gardens, shuns monoculture and factory processing of foods, and stresses the networks of community established when food is produced and consumed locally. Nutrition labels and other nutritional marketing techniques do not take into account these eco-feminist factors.

This feminist perspective acknowledges women's historical expertise about what foods are good. As physicist and eco-feminist Vandana Shiva recognizes in the film *Science for Survival*, women who cook and grow foods with time-tested knowledge and skill know what foods are good because they have a relationship with the people who eat them and they witness daily what makes children and adults healthy, and what makes them sick. Lisa Merton and Alan Dater make a similar point in *Taking Root: The Vision of Wangari Maathai*. Scholars sometimes dismiss as "subjective" or "unscientific" such testimonials by women who grow and cook food as their daily lifetime work, but Keller's argument is that such dismissals reflect a masculine bias. It is not incidental that the so-called "third world" status of the women in these films makes their daily scientific practice even less rigorous in the eyes of the academy.

This process or science in the kitchen involves testing both foods and cooking methods over generations, as well as intuition and a general sense of "goodness" that comes from regular, caring practice. It would be very difficult and expensive, if not impossible, for institutional science to replicate the "science of the kitchen;" for there is no method for lab scientists to monitor daily food-sourcing, cooking, eating, behavior, feelings, and health of large numbers of people over generations. The home "scientists" are able to evaluate the whole person or community in ways that controlled institutionalized science cannot. Of course, institutionalized science can "control" variables in ways the home practitioner cannot. The strengths and weaknesses of each practice constitute another reason that integrating both approaches would facilitate the evolution of the "science" and, in turn, human health.

Some nutrition scientists have endeavored to integrate whole communities, diets, and routines into their studies. Nutrition scientist Kerin O'Dea offers as a more holistic model for nutrition studies. In her model, native peoples have been removed from their habitat and temporarily placed in a location in which only the Western diet is available (In Pollan, 2008, 85-89). These peoples develop modern Western diseases such as diabetes and heart disease, but when they return to their native environment and diet, they recover. Rather than isolating certain nutrients or foods, O'Dea's study examines diet, culture, and lifestyle as a whole. This nutritional approach is consistent with feminist theory, which has traditionally criticized modes of thinking and acting that divide wholes into parts. Strict division of parts reflects a masculine extreme—as Keller says, "Masculine here connotes, as it so often does, autonomy, separation, and distance" (1985, 79).

There is an unwillingness to mix and stir. Thus, even in the "masculine" conceptualization of how to study nutrition, wholes must be divided into parts. Keller emphasizes integration of parts into wholes. This scientific dividing is obvious in nutrition labels.

Attempts to acknowledge the value of women's historical expertise in the kitchen are sometimes dismissed as nostalgic or essentialist. No one is suggesting that women earned their skill and knowledge without hard work. Their work in the kitchen is no more romantic than then scientist's work in the lab. A feminist discourse on nutrition must identify a non-essentialist way to recognize the science women have practiced in the kitchen. To dismiss this recognition as "nostalgic," makes it impossible to have the discussion necessary to integrate kitchen knowledge with institutionalized science. Of course, historically many women have experienced cooking as drudgery and still do, especially when there is no creative freedom involved. Just as valid, however, are the women (and men) like Kalish who develop a skill set and nutritional knowledge base.

Currently, a nutritionally fashionable nutrient is Omega 3 fats. Interestingly, these fats were common in the pre-World War II diet because livestock were sustained primarily through grass, weeds, bugs, and foraging or they drank milk from grass-fed cows. Omega 3 fats are now so trendy that cereal boxes and other packaged foods isolate them, add them to their product, and then market the product as healthy because of its added Omega 3 fats. A feminist theoretical perspective must be skeptical of isolating Omega 3s and placing them in packaged foods: we ask if separating a nutrient from its whole, and then adding it to something else will have the same beneficial effect, as if holistic nourishment were the sum of divided parts. Moreover, ecofeminists would argue that Omega 3 fats are best sourced from traditionally and locally raised or grown foods. As bonus, one is more likely to support a woman farmer when one sources lard or any other food from one's own community.[1]

In another example of isolating parts, nutrition labels prioritize foods with scientifically-discovered nutrients over those that many women practicing in the kitchen before World War II ate and served. This prioritization is problematic because scientists haven't discovered all components of foods and they certainly do not know everything about how these nutrients interact. One example of a food that appears poor on a nutrition label is apples. Nutrition labels report them to contain fiber and a relatively small amount of Vitamin C and Iron. Other nutrients, such as antioxidants and flavonoids are not included on the label. Should we then ignore old knowledge that "an apple a day keeps the doctor a way" and stop eating them in favor of something richer the nutrients that the nutrition label reports? Should we wait until nutrition science can come up with more isolated yet-to-be discovered nutrients in apples? Or should we trust old knowledge? Kalish, describing how her family ate a lot of apples and thrived during the depression years, trusts her own intuition about apples and continues to eat a lot of them (135). Importantly, many recent studies have discovered new

---

[1] Lass says, "there are apparently far more women farmers in the CSA movement than in U.S. agriculture in general" 15.

compounds in apples that affirm what old food wisdom already knew: they are very good.

Institutional nutrition science is incomplete is that it can study only what it knows—so that, for example, it could not examine antioxidants or flavonoids for most of the history of nutrition science because antioxidants had not been discovered. The scientific practice or nourishment at home does not require the discovery, naming, and separating out of individual nutrients.

A feminist approach to food, eating, and health does not wait for nutrition science to discover the currently in vogue nutrients such as Omega 3 fats and antioxidants. Before the rise of food marketing and modern agriculture people ate foods rich in Omega 3 fats and antioxidants without an awareness of the names of these nutrients. However, many women—and men, though with more focus on raising/growing than preparing— have maintained skills and understandings about how to raise/grow, cook, and eat foods that contained these and many other nutrients. Families like Kalish's were confident about what they ate and they saw that their families were healthy and did not suffer from diseases.

Kalish documents that during the Great Depression her family thrived on tried and tested food knowledge. They did so without consulting any institutional science; though interestingly, in retrospect nutrition and fitness science corroborates the science Kalish's family practiced in the farm, garden and kitchen. Producing and eating their own food kept them disease-free as children and adults. With a few exceptions, such as coffee and moderate amounts of sugar, men and women in her family grew and prepared all of their own food. Everyone in her family, regardless of gender, contributed to some form of food production, which is how they survived the Great Depression. Producing food was both gendered and non-gendered: both sexes worked equally and in collaboration to bring food to the table. While the kitchen was women's area of expertise, men were not excluded. As Kalish relates, when there was a big job that had to be done in a limited amount of time in the kitchen, men helped, and vice/versa for outdoor work, such as harvest. Despite the feelings of satisfaction and accomplishment Kalish conveys, she does not romanticize the Great Depression; after all, the title of her book highlights "hard times."

Kalish, like many rural North Americans living prior to the end of World War II, ate whole foods that were traditionally raised or grown on their small, diverse, chemical-free farm. Their milk was full of Omega 3 fats because their cows ate mostly grass, not grain (which is high in the Omega 6 fats), and they didn't pasteurize their milk before drinking it, which protected and preserved the delicate Omega 3 fats and the heat-sensitive Vitamin C. Liberally used dairy fats from their grass-fed cows, such as cream and butter, had high rates of Omega 3s plus a recently discovered fat called conjugated linoleic acid (CLA) which reportedly prevents cancer and heart disease and aids in weight loss (Planck, 68, Pollan, 2006, 267, 431). In addition to the Omega 3s in grass-fed milk (and beef), uncooked milk from grass-fed cows was a good source of Vitamin C—thus, Kalish and her family didn't need to obtain oranges from Florida or Vitamin C tablets to maintain strong immune systems. Nourishing self and family with food grown and raised in forms traditional to humans before World War II was not considered "scientific," as much as it was skillful, resourceful, and nourishing.

Kalish's family did not consult nutrition labels, doctors, or other institutional experts. Nevertheless, Kalish was healthy when she published her memoirs in her nineties. In addition, she and her siblings survived their illnesses with no complications during their youths. When there was an illness, they knew what to do. They possessed similar knowledge and skill regarding what and how to eat. From Kalish's point of view, it almost seems like her grandmother's method for making giblet broth in a way that, coincidentally, prevents it from developing harmful bacteria (bringing it to a "rolling boil" (45)) would be "an old wives' tale," but it is not; rather, it's old knowledge passed down, often, but not exclusively, by and to women. Kalish mentions other tried and tested knowledge about parsnips (which must be eaten in moderation and only when in season) and apples (122, 135).

While usually very confident about old knowledge, as an adult living through the low fat and anti-saturated-fat nutrition "revolution," even Kalish was lured by the "clean," "rational," "scientific" language that condemns animal fats. She describes how she listened to the expert advice to avoid animal fats and substituted a new food, vegetable oil, for the lard in her grandmother's old pie crust recipe. This is an example of rejecting women's tried and tested knowledge for institutionalized nutrition science, which is trendy, often flawed, and not integrated with old knowledge. The nutritional quality of lard, like butter or any animal fat, will depend on the diet and exercise of the animal the fat comes from. The best way to know if one's lard is wholesome is to raise the hog oneself or have an relationship with the farmer who raises the hog, and to have an intimate knowledge of the soil on which it forages and the scraps it is fed.

While nutritionists have labeled lard as unhealthy, lard is similar in profile to the nutritionally-praised olive oil (McCullough, 117). Lard is a traditional staple in the nutritionist-approved Mediterranean diet (though scientists often ignore the prevalence of cured meats and fats as dietary staples) and, contrasted with vegetable oil, has a high smoking point and it does not become easily rancid and therefore carcinogenic (Planck 169). Thus, while contemporary "scientific" knowledge would substantiate Kalish's decision to substitute vegetable oil for lard, a comparison between lard and vegetable oil that takes into consideration a fat's vulnerability to rancidity, cooking temperatures, and fat profile substantiates Kalish's grandmothers' tried and tested wisdom of making pie crust with lard. Importantly, lard, unlike canola oil, can be rendered in one's kitchen. Of course, lard also yields the satisfyingly flaky crust that oil cannot achieve. In the case of lard, "women's old knowledge" is nourishing and delicious.

When choosing between lard and "vegetable oils," a "gender-integrated' conceptualization of nutrition will integrate such factors as the environment and community: good quality, non-hydrogenated lard comes from local pigs that are locally butchered. Lard does not require the processing and long-distance shipping that is energy intensive and sends carbons into the environment. Of course, a high-quality lard that makes it to one's kitchen with minimal ecological impact will depend on an environmentally conscious local farmer. By contrast, vegetable oils such as canola, soybean, and corn are relatively highly processed, often subject to genetic modification and herbicides, shipped at least twice, not available as a local food and not pressed at home. In addition, one must take into account that large monocultures of corn, soybean, and canola contribute to the erosion of soil, and, when not organic, the contamination of water.

The most important questions at stake are not what is always better, vegetable oil or animal fat, regardless of cooking traditions practiced over generations. Rather, a gender-integrated approach emphasizes questions of intimate knowledge about each. What kind of soil did the fat come from? Did my ancestors thrive on this particular fat? What kind of soil did the plant fat come from and what kind of soil did the animal graze or forage on? How was the fat rendered? Were harsh chemicals or harsh temperatures used in the rendering? What effect did the production of this fat have on the soil, water, and air? Who grew or raised the source of this fat? Do I trust this person? Is it even possible to source this fat from someone I have a personal relationship with? Were chemicals applied to the source of the fat? For how long, where, and in what material was the fat stored before I acquired it? What is the smoke-point of the fat? How stable is it? How does it taste? What color is it? What taste and texture will it lend my pie crust, salad, veggie sauté, or pasta sauce?

The list of questions is long and could be longer. There is no end to "intimate knowledge," nor any substitute for it. In all, choosing lard over vegetable oil for the occasional pie crust is a conceptually and politically radical act: it means privileging women's old, but tried and tested wisdom, and choosing a food that's available locally. The industry of food and science has conducted studies that start with the hypothesis that vegetable fats are superior to the animal fats that Americans have traditionally cooked with. What happens when the scientist starts with the hypothesis that the old knowledge of cooking with carefully and intimately produced lard is good and nourishing? This approach is more likely to conceptually integrate the lab with the kitchen.

Science about nutrition must rid itself of the gender paradigms that Keller addresses, and feminists must place relatively newer, industrialized foods such as canola oils, under stronger scrutiny, and less scrutiny on the tried and tested old foods that people ate before rates of heart disease and diabetes and other diet-caused diseases dramatically rose. Feminists cannot rely on a science that rejects women's old tried and tested knowledge about food. As Keller says, science must change, and if we apply her theory to nutrition science, we must integrate old knowledge and give it the benefit of the doubt.

Any call for feminists to recognize and re-value women's historical work with nutrition in the kitchen will undoubtedly be met with accusations of essentialism and confining women to private roles. This author, however, wants to open up nutritional roles for men as well as women in the kitchen. Men and women who are not connected to a tradition of nourishment that pre-dates World War II might research the kinds of foods that women and men were growing and cooking, ask in what ways they were nourishing, and try to pick up those threads of knowledge culturally forgotten.

A gender-integrated conceptualization of nourishment for all peoples is urgent in light of human diet-caused disease epidemics that industrial foods have contributed to. In this new vision, consumers of food for nourishment will seek out harmony and integration between institutional science and the home kitchen.

This vision entails skilled and careful attention to the balance of power between old ways of knowing about food and a practice of institutional science similar to Barbara McClinock's. The other key conceptual transformation will occur between the consumer of food for nourishment and the food itself: the former will know her/his food well—where it comes from, who harvested it, as well as when and how, and many other details. The close relationship between the consumer of food for nourishment and the food itself will involve pleasure, appreciation, humility and respect on the part of the former. Food as nourishment that makes humans grow and live well is a careful miracle for which men and women alike are responsible. The gender-integrated conceptualization of nourishment will be a carefully prepared pot of soup with special attention paid to the balance of flavors and textures. If allowed to simmer gently, the flavors will marry.

References

Berry, Wendell. *The Unsettling of America: Culture and Agriculture*. San Francisco: Sierra Club Books, 1977.

Deutsch, Tracy. "Memories of Mothers in the Kitchen: Local Foods, History, and Women's Work." *Radical History Review*. 110 Spring 2011, 167-177.

Hallberg L, Bjorn-Rasmussen E, Rossander L, Suwanik R. "Iron absorption from Southeast Asian Diets." *American Journal of Clinical Nutrition*. 1877, 30; 539-548.

Hayden, Dolores. *The Grand Domestic Revolution*. Boston: The Massachusetts Institute of Technology, 1981.

Kalish, Mildred Armstrong. *Little Heathens: Hard Times and High Spirits on an Iowa Farm During the Great Depression*. New York: Bantam, 2007.

Keller, Evelyn Fox. *Reflections on Gender and Science*. New Haven: Yale U. Press, 1996.

-- *A Feeling for the Organism: The Life and Work of Barbara McClintock*. New York: Henry Holt Company, 1983.

King-Underwood, Ani. *Science for Survival*. International Broadcasting Trust with TVE, 1995.

Lass, Daniel, and Ashley Bevis, G.W. Stevenson, John Hendrickson, Kathy Ruhf. "Community Supported Agriculture Entering the 21st Century: Results from the 2001 National Survey." Center for Integrated Agricultural Systems (CIAS), College of Agricultural and Life Sciences, University of Wisconsin-Madison. 19 May 2009. <http://www.cias.wisc.edu/wp-content/uploads/2008/07/csa_survey_01.pdf>.

McCullough, Fran. *The Good Fat Cookbook*. New York: Scribner, 2003.

Merton, Lisa and Alan Dater. *Taking Root: The Vision of Wangari Maathai*. Marlboro Productions, *2008*.

Planck, Nina. *Real Food: What to Eat and Why*. New York: Bloomsbury Publishing, 2006.

Pollan, Michael. *In Defense of Food: An Eater's Manifesto*. New York: Penguin Press, 2008.

___. *The Omnivore's Dilemma*. New York: Penguin Press, 2006

Richards, Ellen. *Food Materials and their Adulterations*. Boston: Whitcomb and Barrows, 1896.

Stovall, Holly, Lori Baker-Sperry, Judith Dallinger. "Back to the Kitchen with the Local Food Movement: a Feminist Theory from Farm to Table." National Women's Studies Association Conference, Nov. 11-13, 2011.

Chapter 11: Eating Locally, Staying Healthy: the Perspective of a Gardener and Nurse

Jamie Lane with Holly A. Stovall

I once knew a very interesting lady who was a locavore and lived a sustainable lifestyle on a Wisconsin dairy farm; she tended an organic garden, ate seasonally, canned and froze food for her family to eat year round, and saved her heirloom seeds from year to year. The funny thing is that she never heard the words locavore, sustainable, organic or heirloom. She just did what her mother did, and many generations did before her. She was my grandmother and my first inspiration to start my life as a gardener. Although I only visited her for a week or two each summer, I was completely captivated by her hard work, calm demeanor and her independent lifestyle with grandpa.

Both of my grandparents lived into their upper eighties, living independently in their own home until they died. They rarely went out to eat in restaurants. Grandma cooked their own meat, potatoes and vegetables day after day, assuring that they had strong immune systems. I was told that when they got married, Grandpa wasn't too keen on eating his vegetables. Grandma insisted that he eat his vegetables first while they were hot and he was still hungry before she would let him dive into the meat and potatoes. She probably added years to his life with this simple rule.

As hard as it is for us to imagine now, nearly all gardens were "organic" before the 1940s. And most rural people grew at least part of their own food. This practice was given a boost by World War II and the patriotic victory gardens our parents and grandparents were encouraged to tend. It is unsettling to think that the unjust internment of Japanese farmers in California during WWII helped to promote the local gardening movement…perhaps the only good that came from this dark chapter of American history.

So, what happened to this healthy, self-sufficient spirit we had developed during the war? What happened to all of the backyard gardens and community gardens? Unfortunately, we were sucked in by slick and deceptive advertising, became unmotivated and de-skilled by the convenience of packaged foods and truly believed in the motto of the time: "better living through chemistry." The message was that only the scientific laboratory could produce nutritious food—

not our backyards or our own kitchens. Though there were always skeptics who criticized the industrialization of food, the skeptics were often regarded as conservative radicals who didn't understand science or progress. Most of us knew nothing of the health dangers lurking in packaged foods, but sooner or later, mainstream science would validate the dangers of industrial foods, and the path back to simple home-cooked meals would be slow and challenging; for as a culture, we had forgotten basic gardening and cooking skills. Moreover, as women increased their participation in the workforce, men did not "pitch in" in the kitchen. The burden of cooking remained on women as they entered the paid workforce. Both parties accepted that preparing and eating meals of industrial foods was the "price" to pay for two incomes.

So, why should we grow our own food or buy from local farmers? Let's start with fruits and vegetables. Unless they are grown organically, most fruits and vegetables are exposed to pesticides, herbicides, fungicides and sometimes fecal contamination if livestock is factory-farmed nearby. These contaminants can cause cancers, allergic reactions, reproductive disorders, and food poisoning as well as other problems. They also contaminate the water under the fields which run into our wells and rivers. The popular herbicide Atrazine was banned by the European Union in 2004 due to contamination of the ground water causing reproductive disorders. However, it is still heavily used in the United States.

I am fortunate to be able to garden on an old farmstead that had been abandoned for decades and has a wide safety buffer to the closest corn and soybean fields. It may have been exposed to chemicals in the 50's, but has escaped the recent wonders of modern science in the form of herbicides and pesticides. Not being a soil scientist, I tried to improve my dense clay soil with a variety of additives available to me, including leaves, compost, straw, wood ash and sawdust. I was happy with the outcome and was always able to can and freeze enough food to get us through the winter.

Now my older son is an avid vegetable gardener with more of a scientific tilt. He has studied with local experts and learned of all the nutrients needed for the most productive, healthy soil. His plants are stronger, bushier and more productive than mine ever were. The child is now teaching the parent.

My second son can take the credit for leading me into the world of edible local foods. While I would be out "weeding" my garden, he would be telling me that my weeds were actually very healthy to eat. Soon we were out stalking the wild lamb's quarters, nettles, purslane and wood sorrel while waiting for the spinach to mature. All of these greens are full of necessary vitamins and minerals, not to mention the fact that they are very tasty.

As a health care professional, I know the importance of a strong immune system. It is our first line of defense against diseases. Although it is difficult to separate out all of the chemicals we are exposed to and how they affect us, I can say that in general, my friends who eat a healthy diet are much less likely to catch the flu or develop allergies and autoimmune disorders. A good defense is always the best offense.

Some of these pesticides that I have managed to avoid in the food I grow, but are common elsewhere have also been found to kill off honey bees which are needed to pollinate the foods we grow and, when bees are healthy, can be locally produced in many North-American climates. The so called "colony collapse" disease can mostly be attributed to pesticide toxicity. What did Monsanto do when a research facility found out that Monsanto's neonicotinoids were killing bees? Of course, they bought the research facility. Do you think Monsanto's studies will come up with the same findings?

Each spring I am very relieved to find that my bees have survived another winter. I attribute this to three causes. First, the bees have a lot of uncontaminated garden as well as prairie flowers to feast on. They would have to travel a ways to be contaminated by agricultural chemicals. Secondly, I always leave the bees a lot more honey to get through the winter than is recommend by the experts. I would rather leave them a surplus than need to feed them sugar water to keep from starving in the spring. Many commercial honeybee keepers feed their bees high fructose corn syrup in March and April. This is made from corn sprayed with the neonicotinoids that are killing the bees in the first place! The third reason my bees survive can be called proper neglect.

Besides the delicious taste of local honey, there are health benefits as well. Eating honey containing local pollen can help decrease reactions to local allergens. This process works similar to a vaccination. The small amount of pollen in the honey can trigger an immune response against the pollen spores so that a person with allergies has a better defense against the real pollen sources. There are no peer-reviewed scientific studies to prove this effect, but there is enough logical and anecdotal evidence to make it worth a try. Moreover, because honey has a strong flavor, people don't usually eat too much of it, so it does no harm.

Another health reason to eat local produce is that you can meet your farmers and know what farming techniques they use, such as what is added to the soil to enrich the vegetables. According to Dr. Max Gerson, "Soil must be regarded as our external metabolism which produces nutrients for our internal metabolism." The line from soil to vegetable to our mouths is direct and clear, and most commercial soils have exhausted their supply of necessary minerals as is evident from the quotes that follow:

According to Rex Beach, in a document considered by the 74th Congress in 1936, "… 99 percent of the American people are deficient in … minerals, and … a marked deficiency in any one of the more important minerals actually results in disease." If mineral deficiencies in the 1930s were caused by neglecting to properly care for the soil, this problem can only have gotten worse; for modern farming methods are much harder on the soil. That modern farming practices causes soil erosion is well-documented. Beach continues, "Do you know that most of us today are suffering from certain dangerous diet deficiencies which cannot be remedied until the depleted soils from which our foods come are brought into proper mineral balance? The alarming fact is that foods–fruits and vegetables and grains–now being raised on millions of acres of land that no longer contain enough of certain needed minerals, are starving us–no matter how much of them we eat!" One can be sure of the quality of the soil one's food comes from only if one has a "close relationship" with that soil. Currently, most produce sold in the Midwest comes from California, making it impossible for Midwestern consumers of food to know the soil their foods come from.

Most commercial farms only add nitrogen, potassium and phosphorus to the soil. There are many other micronutrients that are needed for healthy food, such as zinc, boron, copper, iron and manganese. One source of soil enrichment is aged animal manure. The mineral value will depend on the feed for the animals. Compost is another source of nutrients. This can take months to age, depending on how it is managed. (Compost also loosens the soil and provides better drainage.) The best way to know what minerals your soil is lacking is to have it tested, and to purchase those you are lacking. The flavor and nutrition of fruits and vegetables are more dependent on the mineral composition of the soil than on its organic matter. I am thankful that my son has taught me about healthy sources of soil additives, such as sphagnum peat, blood meal, Green Sand and Azomite.

It goes without saying that eating local provides fresher foods. Fruits and vegetables lose some of their nutrients when washed, chopped or heated, so the sooner you can eat an item after picking it, the better the value will be. If the food is shipped from California, Chile, or even China, the nutrients will be greatly diminished. This fact doesn't even take into consideration the amount of fuel burned by shipping the food across the country, and all of the air and noise pollution from the trucks' engines.

There is no substitute for the convenience and spontaneity of being able to source food from the "backyard," or from my cupboard of home-canned goods. The food does not travel—not even from the store, much less from far-away soils. As stated, the food freshly picked form the garden is delicious. It contributes color, texture, and flavor to the table. I chose fresh herbs to vary the flavor. While I might, for example, source beans or pasta from the grocery store, I surround them with just-picked vegetables or homemade tomato sauce. It makes a wonderful meal full of flavor and nutrients.

While I'm on the subject of vitamins and minerals, each one has a specific role to play in our bodies. For example, vitamin C is good for our tissues and Vitamin D is good for our bones. When a vitamin is missing from our diet, we can call the problem by its medical term, such as scurvy, or we can simply call it Vitamin C deficiency. For Vitamin D, we can call it rickets or Vitamin D deficiency. I believe this holds true for most diseases. The basis for almost all illnesses is due to either a *deficiency* of some vitamin, mineral, enzyme, bacteria or other naturally occurring part of our physiology, or it is due to some type of toxicity inflicted on our bodies. In other words, we have too little of something we can't live without or too much of something that we can't live with. Eating foods from soil we know is healthy helps assure we get balanced nutrients in our diet.

If we put the unseen nutrition content of fruits and vegetables aside and deal with the obvious, the taste and texture of fresh foods is astounding. It is a joy to watch someone bite into a local, fresh tomato or carrot for the first time. So much produce that we buy from the grocery store has been hybridized and selected for its ability to retain its shape and color while packed into crates and shipped around the world. Taste is not a priority for the producers, so we are stuck buying their beautiful but cardboard-flavored tomatoes. The moisture content can't compare to fresh either. A carrot straight from the ground practically explodes with carrot juice when you bite it to it.

When it comes to meat, there are once again several good reasons to know its source, and the best way to know the meat's source is to know its producer. Feedlot owners regularly feed antibiotics to cattle to prevent infections, which in turn leads to antibiotic resistance in humans. The Union of Concerned Scientists estimates that 70% of the antibiotics used in the United States are used for livestock. The cattle are also fed estrogen to fatten them up, which can lead to hormone imbalance in humans whether we are eating the meat or drinking the milk. Thirdly, grass-fed cattle produce healthy omega 3 fats (which are linked to prevention of several brain and heart diseases) in their tissues, which is quickly lost when cattle are moves to a feedlot for their last months of fattening up on corn. Grass fed cattle have two to four times as much omega 3 fats as feedlot cattle.

As for hogs and chickens, they are also routinely given antibiotics to prevent infections. Chickens that are grass fed can have 20 times more omega 3 fats in their meat and their eggs compared with confined chickens. The egg yolks are also darker yellow if the hens are grass fed. People who procure local eggs often notice that the yolks are darker than the pale ones from the grocery store: chickens that spend their days on pasture get more exercise and a healthier diet. Commercial eggs, even those that boast they come from chickens that are "free range," probably do not have access to pasture. Technically, to qualify as "free range" the chickens only have to have access to a small, concrete patio.

Another healthy aspect of local eating we often overlook is how to maintain a good balance of beneficial bacteria. This has been in the news frequently in the past year or two. Scientists have discovered that over sixty percent of our immune system is controlled by our bacteria in our colon, and if we develop the wrong types or amounts of bacteria from a bad diet or from antibiotics, it can cause negative, unforeseen consequences. Dr Natasha Campbell-McBride, a British neurologist, has discovered that autistic children have abnormal colon bacteria, as do their mothers, most likely from excessive antibiotic usage over two or three generations.

One way to replenish missing good bacteria is to eat unsweetened yogurt with live yeast cultures. It is always tastiest if you make your own yogurt from fresh, local milk. Another way is to eat fermented vegetables such as sauerkraut or kimchi, fermented Korean cabbage. Sauerkraut is easy to make—all it takes is cabbage, salt and water. When these cultured or fermented foods are made at home from local foods, they are nutritionally superior to their counterparts in the grocery store: one has more control over the nutritional components of the raw ingredients, and just as importantly, one has control over the additives; for much of the commercial cultured or fermented foods have added preservatives, sugars, and artificial flavors. Indeed, commercial yogurt is likely to have as much added sugar as a candy bar. It is easy to "unjunk" yogurt: take plain yogurt and add local honey or preserves from local fruit grown in nutrient rich soil. It must be added that culturing and fermenting your own foods takes very little hands-on time: the process is mostly about leaving the food alone and letting the bacteria do its work.

What are our choices if we don't grow our own food? One option is to buy local food from a farmers' market. I love to go from booth to booth talking to each farmer about their successes and frustrations each week. When I think about how early they had to rise to have fresh produce ready for me by 7 am, I am reminded of how hard they work for each dollar. In turn, I appreciate the true costs of good food: good food is not cheap and should not be cheap, for the farmer's market vendor worked hard and long hours to cultivate that food, harvest it by hand, keep the food's soil healthy, bring it into town, and spend their Saturday mornings, when everyone else has off of work, selling their goods.

Another option for local food is to subscribe to community supported agriculture (CSA). When you join a CSA, you pay a certain amount (approximately $20.00 per week) and, for example, can pick the amount and types of vegetables, fruits and flowers listed on the chalk board for that week. However, a CSA can provide much more than great local food. A CSA can become a community of like-minded citizens who get together to socialize each week while their children run around on the farm and get a healthy dose of dirt and exercise. It can become the family you always wish that you had, but never did. This community is especially important in today's mobile workforce, which requires us to relocate to places far from "home."

If we don't have a farmer's market or CSA, we can buy some relatively fresh produce at the grocery store. Some food stores are even starting to carry seasonal local produce, but this is a small percentage of the entire department. Most people also consume canned, frozen or packaged food from the grocery store. Buying packaged fruits and vegetables can (and should) lead to an endless detective game of finding the hidden harmful ingredients. Harmful additives can be something as simple as too much sodium, which raises one's blood pressure, or too much sugar and trans fats which can cause diabetes and heart disease. Some packaged foods contain MSG or aspartame, both being flavor enhancers which are glutamates and cause dangerous over- stimulation of neurotransmitters in the brain. Then there are the preservatives, stabilizers and food colorings which can cause many various reactions. According to Florence Williams, marketed food is often packaged in BPS-containing plastic. BPA is an estrogen mimic that has been linked to hormone imbalances that cause cancer and other chronic health problems. Even your grocery store receipt contains BPA.

We would all be a lot safer if we didn't buy anything with packaging or a nutrition label. Moreover, we should *avoid* products that tout health claims: as nutritionist and author Marion Nestle says, "From the ivory tower in which I sit, the remedy is easy, don't allow health claims on processed foods at all. The claims are all inherently misleading, as would be obvious if you gave it a minute's thought" (2011). Food, says Nestle, is not a drug and should not advertise health benefits. Many products that do claim health benefits, such as highly processed cereals and yogurts, often exaggerate the positive nutrients and downplay the sugars and trans fats. I like the saying "if your grandmother wouldn't recognize it, don't eat it." Highly processed foods with added sugars and transfats were simply not widely available or affordable to the figurative "grandmothers" who lived much of their lives before World War II.

According to the National Gardening Association survey done in 2009, over one third of all Americans are growing some of their own food: 58% said that they garden to get better tasting food, 54% to save money, 51% to grow better quality food and 48% to be sure about the safety of their food. No matter the reason that people garden, it provides healthy exercise outdoors, often with other family members. The satisfaction of growing one's own vegetables is immeasurable and links us by an invisible bond to our ancestors. Thanks for the inspiration, Grandma.

References

Aro, A., S. Mannisto, I. Salminen, M. L. Ovaskainen, V. Kataja, and M. Uusitupa. "Inverse Association between Dietary and Serum Conjugated Linoleic Acid and Risk of Breast Cancer in Postmenopausal Women." *Nutr Cancer* 38, no. 2 (2000): 151-7.

Beach, Rex. "Modern Miracle Men: Relating To Proper Food Mineral Balances by Dr. Charles Northen." 1936. Web. 27 Feb. 2013. http://www.senatedocument264.com/.

Campbell-McBride, Natasha. *Gut and Psychology Syndrome. Natural Treatment Of Autism, ADHD, Dyslexia, Dyspraxia, Depression And Schizophrenia. Cambridge:* Medinform Publishing, 2010.

Colangelo, Aaron, and Jennifer Beth Sass. "European Union Bans Atrazine, While the United States Negotiates Continued Use," *International Journal of Occupational and Environmental Health.* Vol 12/No 3, Jul/Sep 2006, 260-267.

Dhiman, T. R., G. R. Anand, *et al.* (1999). "Conjugated linoleic acid content of milk from cows fed different diets." J Dairy Sci 82(10): 2146-56.

Ferrie, Helke. "Healing – Not Drugging! The Gerson Therapy for Cancer Sixty years of Proven Success." *Rawfoodinfo.com.* Rawfoodinfo.com. Nov. 2004. Web. 27 Feb. 2013.
http://www.rawfoodinfo.com/articles/art_gersontherapy.html.

Harris , Gardiner. "U.S. Tightens Rules on Antibiotics Use for Livestock." *NYT.* 11 April, 2012. Web. 27 Feb. 2013.

Gucciardi, Anthony. "Blamed for Bee Collapse, Monsanto Buys Leading Bee Research Firm." Natural Society. Naturalsociety.com. 19 April, 2012. Web. 27 Feb. 2013. http://naturalsociety.com/monsanto-bee-collapse-buys-bee-research-firm/#sthash.Ys01Ul8q.dpuf

Ip, C, J.A. Scimeca, *et al.* (1994) "Conjugated linoleic acid. A powerful anti-carcinogen from animal fat sources." p. 1053. Cancer 74(3 suppl):1050-4.

National Gardening Association. "Survey Reveals Double-Digit Increase in Food Gardening for 2009: National Gardening Association poll finds more American households are starting food gardens to "grow their own groceries." Feb. 26, 2009. Web. 27 Feb. 2013.
http://scotts.mediaroom.com/index.php?s =43&item=66

Nestle, Marion. "Foods with Benefits: Oh, please." Food Politics. Foodpolitics.com. 15 May 2011. Web. 27 Feb. 2013.
http://www.foodpolitics.com/2011/05/foods-with-benefits-oh-please/.

Rose, D. P., J. M. Connolly, *et al.* (1995). "Influence of Diets Containing Eicosapentaenoic or Docasahexaenoic Acid on Growth and Metastasis of Breast Cancer Cells in Nude Mice." Journal of the National Cancer Institute 87(8): 587-92.

Rule, D. C., K. S. Brought on, S. M. Shellito, and G. Maiorano. "Comparison of Muscle Fatty Acid Profiles and Cholesterol Concentrations of Bison, Beef Cattle, Elk, and Chicken." J Anim Sci 80, no. 5 (2002): 1202-11.

Simopolous, A. P. and Jo Robinson. *The Omega Diet*. New York, HarperCollins, 1999.

Union of Concerned Scientist. “Our Failing Food System.” ucsusa.org. 30 Aug. 2012. Web. 27 Feb. 2013. http://www.ucsusa.org/food_and_agriculture/our-failing-food-system/

Williams, Florence. *Eat Like a Mennonite*. *New York Times*. 18 Jan. 2013. Web. 27 Feb. 2013. http://www.nytimes.com/2013/01/19/opinion/eat-like-a-mennonite.html.

Chapter 12: The Birth of the Macomb Community Garden

Heather McIlvaine-Newsad and Rob Porter

The Roots of the Movement

When my husband and I moved to Macomb in the summer of 2000, we were excited to be returning to the Midwest. Both born and raised in southern Ohio, we had spent the last five years in Florida where the soil was sand and things grew in the winter and decomposed in the summer. I knew from growing up on a farm that the soils in Illinois were among the best in the world and we were excited to have a garden of our own where we could grow flowers, tomatoes, and maybe some greens.

We were also fortunate to be moving to a community where we had friends. Not all academics have this luxury. John Curtis and his wife Karen Mauldin Curtis and I were Peace Corps Volunteers together in the Dominican Republic many moons prior. John had been assigned to an abandoned USAID organic model farm a couple of mountain ranges over from the community where I lived. While I worked on reforesting a watershed, he learned alongside those in his village how to farm without the expensive and harmful chemical inputs. It was backbreaking and palm skinning work, but after two years he had transformed the abandoned hillside into a real farm that the community members took pride in. He fell in love with farming and with the idea of being a farmer. In 2001, when John and Karen started their community supported agriculture farm (CSA), *Barefoot Gardens*, just outside of Macomb, IL we were among the first happy subscribers.

Barefoot Gardens is different from other CSAs I have researched. It shares the philosophical underpinnings with other farms in the movement, which is "to bring people, food, and land together by creating a partnership between community members and local food growers" (Barefoot Gardens 2012). However, *Barefoot Gardens* is more than a weekly box of vegetables. People come to the farm not only for the vegetables, but to socialize, watch their children run around exploring nature, and share a glass of wine and conversation. Although John continues to grow wonderful, healthy food, *for me* the most important thing he has cultivated is community.

The community *Barefoot Gardens* has created and nurtured is ethnically diverse. This is no small feat in rural Illinois. The people are also highly educated, and solidly upper- middle class. The members of this community have the discretionary income to be able to pay for the produce before receiving it. Early in the history of the garden, John, Karen, and many of the members of the CSA talked in depth about how the CSA model could be made more inclusive, so that those without the financial means to join the garden could benefit as existing members did. The basic economic structure of a CSA, based on the upfront payment of a share price, means that what is grown in the gardens are not available to everyone. However, *Barefoot Gardens* has gone out of its way to be as inclusive as possible by creating scholarships, fashioned sliding fee scales, and sometimes bartered their produce for other skills, and yet many in the community who would like to be able to participate in the garden are not able to. Even though compared to large-scale organic produce one sees in grocery stores, *Barefoot Gardens* is competitive in price. At the root of the problem is that most organic produce for sale in the United States is out of reach of most low-income households.

Over the years that these conversations were taking place, our families and the gardens were growing. Other changes were happening too. Two conventional farms on either side of *Barefoot Gardens* installed CAFOs (confined animal feeding operations) raising thousands of hogs destined for a meat packing plant owned by Cargill in nearby Beardstown. The two-lane blacktop road to the garden was soon crowded with semi-trucks driving at breakneck speed and the air was filled with the stench of thousands of hogs housed in close quarters. *Barefoot Gardens* and the house that John and Karen had built themselves were put up for sale. The Curtis's and the gardens moved and found a new home closer to Macomb.

From my perspective the move has been both good and bad. Their client base has expanded now that they are closer to town. From an economic standpoint, the relocation has been excellent. However, the change altered the community for me. Prior to the move and expansion, I felt like the members of the garden were part of my family. Afterward, seeing many faces I didn't know just annoyed me. Added to this was the crazy schedule of my busy girls. Gone were the carefree days when I could go to the garden and hang out for a couple of hours not worrying about having to pick up one child from swim practice or take another one to karate. I could enjoy the space and the people. I could "be" in the garden. Now, having to schedule a garden visit into an already packed week seemed like a chore rather than a pleasure. Added to these variables were the nagging questions of equity and food democracy in the CSA model. I focused on forming a new garden, one that would be accessible to all people regardless of income level. While I knew that tending my own garden would be more labor and time intensive than picking up produce at the CSA, I was also drawn to the concept of being able to be in control of my time in the garden. If I wanted to water at five in the morning before my run, then I could do so and swing by after I was done to turn off the water and pick some fresh flowers. I also understood that my failure to tend my plot might result in lots of weeds or fewer carrots, but that was okay too, since I was in control. So, when Rob Porter arrived as a new faculty member in the Recreation, Parks, and Tourism Administration (RPTA) department in 2009, it seemed like an opportune time to create a space that would be accessible to a wide range of people in Macomb. At this point, the *Macomb Community Gardens* (MCG) was born.

The Long Labor and Birth

The birth was not an easy one. In fact the labor lasted almost two years. The first big hurdle was land. Neither Rob nor I had any land, and we had not cultivated any local political connections to be granted access to land. We were what anthropologist John Bodley (2007) might call "landless peasants." So, we started with the community we were members of, Western Illinois University (WIU).

As part of the RPTA department, Rob has access to Horn Field Campus, 92 acres of land located one mile south of Macomb. The site contains woodlands, prairie, and some ground that is suitable for gardening. In the fall of 2009 Rob worked with some graduate students on the site to create the first iteration of the MCG. While it produced some vegetables the following spring, the garden generally failed because of its late start. As well, the growing season was wet; the site did not drain well, and once the graduate students left no one tended the plots. To compound this problem, Horn Field Campus is not well known outside of the RPTA department so no new students came in and took over. Many WIU students go through their entire undergraduate careers without every setting foot on that part of the campus. In addition, as we began to think about the future of the garden, hypotheticals began to surface: if the garden prospered, would non-WIU members have access to the garden? And if someone was injured at the garden, who was liable? While this first garden might have become functional in time, these initial factors led to its early demise.

As a member of the RPTA department, Rob had been involved with the Macomb Park District and knew many of the board members. Recognizing that the garden needed a new location, Rob knew who he had to talk to in the community in order to acquire access to green space to secure a garden plot during winter 2010. This plot was on city park land, a newly constructed sports facility including soccer and softball fields. The land is owned by the Macomb Park District and they were thrilled at the prospect of adding a community garden to the setting.

The softball and soccer fields are located along the northeastern border of Macomb, and although they are technically within the city limits, they are bordered by corn and soybean fields, woods, and industry. They are not accessible via public transportation and the nearest house is a mile away. We began to spread the word that there would be a new location for the MCG. We used word of mouth, talking to people in the university community who loved to garden, but more importantly letting members of the Macomb community for whom a CSA was not option know about the garden. We held three information meetings at the Macomb Town Hall, advertised widely in the local paper and also recorded a couple of radio spots on our local NPR and commercial radio stations. In the end, we had between eight to ten families. We also hosted hundreds of deer, experienced one of the wettest springs ever, and without the drainage trench digging abilities of Joel Gruver (an agriculture professor at WIU also involved in the gardens) we would have produced nothing.

However, these challenges did not hamper our enthusiasm for the project. We went back to the literature and worked on our local political connections at the same time. Knowing the social and economic value that community gardens bring to urban settings by transforming dilapidated and vacant lots into usable spaces (Landman 1993, Linn 1999, Pottharst 1999, Glover 2004), we set out to find such a space in rural Macomb. In fall 2011, Rob learned that the new softball fields on the city park land would leave the existing fields vacant. The fields about to be abandoned were on Housing Authority of McDonough County (HAMC) land, U.S. Department of Housing and Urban Development (HUD) funded public housing. Additionally, there was talk of expansion of the soccer facilities onto the land where the garden was located, so the garden would have to be relocated at some point in the future. Rob immediately approached Bill Jacobs, the director of the HAMC and asked if we could use the newly abandoned softball field on HAMC property. Bill loved the idea of having a community garden next to HAMC housing. Locating the gardens on HAMC property, next to low income housing would help low-income populations gain access to the environmental "good" (Salazar 1998, Porter 2009) of green space for gardening. In turn, this addressed the concerns we had about gaining access for *every citizen* to the healthy food that comes from gardening. In typical fashion, we plowed ahead with little money and much enthusiasm.

In 2011 the garden broke ground for a third (and hopefully final) time in a location that is logistically much better suited to long term sustainability and success. MCG is now located on a public bus route, is part of a neighborhood, has access to city water for irrigation, and is within walking distance of the local elementary school. During the 2012 growing season the garden was composed of approximately 85 families growing food on over 130 plots.

Food Democracy

What started as a project to allow people space to grow their own food, has blossomed into something that is much more complex and robust than we could have ever anticipated. We both share concerns about the globalization of our food system, but we are also concerned with how we could make the community of Macomb a better place for all its residents. From this perspective the community garden is less about growing food and more about giving people a "third place" (e.g., coffee house, local pub, church) outside of work and home (Oldenburg 1999) where they can gather, network, and identify together as part of a community (Linn 1999, Moncrief & Langsenkamp 1976, Schrieber 1998, Glover 2004)."

On the surface Macomb appears to be a homogeneous rural community. Upon closer inspection it is a microcosm of the changing face of rural America and is full of extremes. It is located in McDonough County in the western part of Illinois. WIU and its roughly 12,500 students is a focal point of the town. Yet, McDonough County has the highest percentage of poverty when compared to contiguous counties. Nearly 21% of residents of McDonough County are below the federal poverty line as compared to 12.2% of US residents. Twenty-two percent of the children in McDonough County live below the poverty level (U.S. Census Bureau 2008). Economically, the region has suffered for decades, beginning in the 1990s when manufacturing plants began to relocate their businesses overseas. Approximately five factories that once contributed to the area's employment are no longer in existence. Outside of WIU, which employs over 1,800 people, the other main economic activity is farming. The soils in this region are among the most fertile in the United States. The farms in the area reflect the national trend in industrial agriculture with a decreasing number of farms, while the size of farms increases (McIlvaine-Newsad et. al 2004). The crops produced in the area are not food, but commodities like corn and soybeans. McMichael (2000:23) writes,

> In the US, two percent of the farms grow fifty percent of agricultural produce, the average family farm earns only fourteen percent of its income from the farm, and ninety-five percent of American food is a corporate product. The food industry is the largest American industrial sector, but it doesn't produce food security, as thirty million Americans are hungry.

Many of the early members of the community gardens have personal and professional concerns about the current state of our food system. After several years of commentaries on the local NPR station, letters to the editors, and participation in the town's "Heritage Day's Parade" that promoted "grow your own," more people in the community are becoming aware of the problems with the way we grow and process our food. All of the community gardeners subscribe to the mantra "eat local and seasonally" and routinely chastise each other in a friendly nature when caught at the local grocery store buying some vegetable "out of season."

As Michael Pollen writes, "monoculture is at the root of virtually every problem that bedevils the modern farmer" (2002:225). Nowhere is this more evident than where we live. Several years ago corn blight hit the region and thousands upon thousands of acres of corn toppled over making it very difficult to harvest. The community gardens with their ecological diversity provide a safeguard against such diseases wiping out entire crops. We look at the community in a similar fashion. The more diverse our community is the more robust and resilient it will be. Perhaps equally important has been the notion that the gardens have fostered the expansion of local sustainability as mentioned by Holland (2004). Building on work by Stocker and Barnett (1998), Holland writes that community gardens are "agents of change" in the way in which community members create and manage community spaces, and disseminate community science and innovative technologies. Gardeners routinely learn from each other as they chat while working on their individual plots. Newly discovered heirloom seeds are exchanged along with recipes for organically based pesticides and zucchini bread, and advice is shared on how to best insulate the pipes in your house so they won't freeze in the winter.

Whether they recognize it or not, the MCG gardeners are practicing "food democracy." In the mid-1990s Tim Lang (1999) developed the term food democracy as a response to the increasing corporate control and lack of consumer participation in the food system[1]. Food democracy is based on the principle that citizens or "food citizens" have the power to determine food policies and practices locally, regionally, nationally and globally. Food democracy asserts that it is a right and responsibility of citizens to participate in decisions concerning their food system. These beliefs challenge the corporate structure and allow for bottom-up control of the food system. This process transforms individuals from "passive consumers into active, educated citizens". The word "democracy" comes from Greek words meaning "people" and "rule." How, then, do we ensure that "the people" rule our food system? The goal of food democracy is to ensure all citizens have access to affordable, healthy, and culturally appropriate foods. Food democracy emphasizes social justice in the food system, and food is viewed as the center of the democratic process. In the following section, we will offer two examples of how the MCG practices food democracy.

Community Plots

One of the unique aspects of the MCG is the "community plots" that Joel Gruver oversees. Over the last two years large plots containing a wide variety of potatoes, sweet potatoes, onions, garlic, corn, asparagus, pumpkins, sunflowers, strawberries, raspberries, and blackberries have been planted, tended and harvested for all members of the MCG to share.

The seeds or starts come from various sources. We often rely on Ann Runner, a FFA (Future Farmers of America) teacher at West Prairie High School to have her class start some of the plants. Other times we take clipping from our members' backyard gardens for use in the community plots. If we have to buy the seeds, we use the community funds that are generated from the rent members pay each year ($20 per plot and $10 for each additional plot) to garden. Joel periodically sends out emails organizing the MCG members to do a massive weeding, mounding of potatoes, harvesting, or similar work. Generally between 10-15 gardeners show up and there are usually lots of children in the mix. The rule of thumb is, if you have contributed to the upkeep of the community plot you can take your share from what is harvested. This is an aspect of community gardening that is unique and not found in many other community gardens around the world.

Sharing the Harvest

Healthy food costs money and the most expensive food in the grocery stores are fruits and vegetables. In McDonough County 23.4% of the residents below the federal poverty line (U.S. Census Bureau, 2008). In our small community we have four food banks, one monthly soup dinner, and a weekly Friday after school program that gives children between the ages of 5-13 dinners and a backpack full of food and books to take home over the weekend.

Public health statistics tell us only 17.9% of the McDonough County adult population eats the recommended daily intake of five servings of fruits or

[1] Hassanein (2003), Shiva (2005), and Hamilton (2005) have furthered this concept.

vegetables. Not surprisingly, 23.2% of the adult population of the county is obese (Illinois Department of Public Health 2003). The average income of HAMC residents is $12,458 and the obesity rate is particularly high among HAMC residents. The rural poor in our community are facing a double-edged food problem: how to be both food secure and healthy. As such, many of the gardeners found meaning in serving others by distributing food to other gardeners, HAMC residents not involved in the garden, and local food banks. One gardener from the HAMC commented,"We have a great number of poor in this small community, you know, and they can use that stuff." Another describes how she organized HAMC residents into a food distribution initiative for HAMC residents not working at the garden.

> "We go out and pick and usually I would try to get a couple people out there with me, sometimes we'd have five or six, it just depended. Then we'd come in to the kitchen here and separate it into bags, you know lettuce... and then basically we have a distribution list of everybody that's signed up and what type of vegetable they've signed up for. So we would just go down the list after we've bagged the vegetables and mark off each person what they were taking on the list, and then put it all together and took it up to each apartment and hung it on their door handle."

As a member of the board of the largest food bank in Macomb, Heather would routinely take in bags of community potatoes, onions, and tomatoes. Food banks in our region rarely stock fresh vegetables and anything brought in the morning would be gone in the afternoon. Another gardener, Melissa Calhoun installed a green table next to the tool shed at the MCG site. Gardeners who have excess produce can put anything they would like to donate on the table and Melissa picks it up and distributes it to local food banks.

It seems clear to us that the creation of community plots and the donation of food to those in need are indeed food democracy in action. The gardens have quickly developed into a vehicle for social change (Alaimo et al. 2010). Being able to engage in healthy physical recreation, decompress mentally, and grow your own vegetables not only results in a healthier, less obese community, it also *counteracts* the physical, economic, and social negatives created by the current corporate approach to food production and provision. Thus, by engaging in community gardening, local citizens are not only providing themselves with fresh, organic, affordable produce. They are making a conscious decision to resist large-scale agriculture and its dependence on environmentally unsafe practices such as heavy chemical use and petroleum-dependent transportation. As Wendell Berry (1981) states:

> "We like the thought that the outdoor work that improves our health should produce food of excellent quality that, in turn, also improves and safeguards our health....Gardening--or the best kind of gardening--is a *complete* action."

At this point the community gardens have brought together people whose paths might not cross on a daily basis -- low-income HMAC residents, professors, and factory workers. One of the gardeners commented directly on the economic and educational diversity brought together by the garden.

> "I got to meet people I never would have met before. I'd never met you and "Greg" or, or "James" or any of you guys if I hadn't been in that garden project. I was out there visiting with the other gardeners and I enjoyed that more than anything because these were people I never would have met before. You know, I look forward to meeting more people from the community... and talking to them and see what they think about life and the world and their garden and, you know, and sharing. That's what, that's what it's all about, sharing."

The word *community* comes from *communis* "common, public, general, shared by all or many," (OED 2012). The syllables of *common* come from separate roots, the first meaning "together" and the second meaning "next to" (Sanders 2001). This is the story of *Barefoot* Gardens and the *Macomb Community Gardens*. They have grown up together and next to each other. I remember having many conversations with John Curtis about how he would measure the success of his CSA. Would it be through the amount of income generated or the pounds of tomatoes grown, or the number of members enrolled each year? None of these answers were what John answered. As he writes on his website:

> "I believe strongly in the importance of local food and no food is more local than the food you raise for yourself, your friends and/or your family. I know that people want to learn to garden." (Barefoot Gardens 2013).

When I left *Barefoot Gardens* as a member, there were many factors contributing to my departure. But I never could have begun to create a community garden had it not been for what I had learned from John. The community gardens project set out to reclaim unused space allowing individuals from rural Illinois the opportunity to grow their own vegetables. The result, in addition to tons of tomatoes and spinach and sweet potatoes, is a group of disparate individuals, some former members of the CSA others having never gardened before are growing a vibrant community. Underlying concerns about an unsustainable food system, hunger, food insecurity, obesity, and their love of gardening brought them together; community and food democracy keep them together. As Wendell Berry (1985) put it, "What I stand for is what I stand on."

References

Alaimo, K, Reischl, T., & Ober Allan, J. Community Gardening, Neighborhood Meetings,and Social Capital. *Journal of Community Psychology* 38, no. 4 (2010): 497-514.

Barefoot Gardens. 2012. http://www.barefootgardens.org/about-us.html. Accessed 12/18/2012.

Berry, W. *Collected Poems*. New York, NY: North Point Press, 1985.

Berry, W. *The Gift of Good Land*. New York, NY: North Point Press, 1981.

Bodley, J. *Anthropology and Contemporary Human Problems*. New York, NY: AltaMira Press, 2007.

Glover, T. Social Capital in the Lived Experiences of Community Gardeners. *Leisure Sciences* 26, (2004): 143-162.

Hamilton, N. Moving Toward Food Democracy: Better Food, New Farmers, and the Myth of Feeding the World. *Drake Journal of Agricultural Law* 16, (2011): 11-29.

Hassanein, N. Practicing food democracy: A Pragmatic Politics of Transformation. *Journal of Rural Studies* 19, (2003): 77-86.

Holland, L. Diversity and Connections in Community Gardens: A Contribution to Local Sustainability. *Local Environment* 9, no. 3 (2004): 285-305.

Illinois Department of Public Health. *IPLAN Data Set: McDonough County Obesity*. http://app.idph.state.il.us/Accessed October 12, 2010.

Landman, R. H. *Creating Community in the City: Cooperatives and Community Gardens in Washington, D.C.* Westport, CT: Bergin & Garvey, 1993.

Lang, T. *Food Policy for the 21st Century: Can It Be Both Radical and Reasonable?* in Koc, et. Al (eds). For Hunger-Proof Cities: Sustainable Urban Food Systems. (216-224) International Development Research Centre, Ottawa, Canada.

Linn, K. Reclaiming the Sacred Commons. *New Village,* 1, no. 1 (1999): 42–49.

Mcilvaine-Newsad, H. et al. Direct from Farm to Table: Community Sponsored Agriculture in Western Illinois. *Culture and Agriculture.* 26, no. 1-2 (2004): 149-163.

McMichael, P. The Power of Food. *Agriculture and Human Values* 17, (2000): 21-33.

Moncrief, L.W. & Langsenkamp, R. Cultivating Community Gardening. *Parks and Recreation,11,* no. 4 (1976): 19–21, 39–42.

Pollan, M. *The Botany of Desire*. New York: Random House.

Oldenburg, R. *The Great Good Place: Cafes, Coffee Shops, Bookstores, Bars, Hair Salons, and Other Hangouts at the Heart of a Community*, 3rd Ed. New York: Marlowe & Co, 1999.

Online Etymology Dictionary. 2012. http://www.etymonline.com/index.php?term=community. Accessed February 27, 2013.

Porter, R. *Environmental Justice and North Georgia Wilderness: A GIS Based Analysis.* Saarburcken, Germany: VDM Publishing, 2009.

Pottharst, K. 1995. Urban Dwellers and Vacant Lots: Partners in Pride. *Parks and Recreation, 30*, no. 9 (1995): 94–101.

Salazar, D. Environmental Justice and Natural Resource Management in the Pacific

Northwest. *Northwest Science Forum* 72, no. 1 (1998): 52-57.

Sandres, S. The Common Life. E. Freyfogle (ed), *The New Agrarianism: Land, Culture, and the Community of Life.* (pp 217-238). Washington: Island Press, 2001.

Schrieber, P. Community Gardening: Design, Techniques, and Tools. S. P. Simson & M. C. Straus (eds.), *Horticulture as Therapy: Principles and Practice* (pp. 377–397). New York: The Food Products Press/Haworth Press, 1998.

Shiva, V. Earth Democracy: Justice, Sustainability and Peace. Cambridge, MA: South End Press, 2005.

Stocker, L. & Barnett, K. The Significance and Praxis of Community-Based Sustainability Projects: Community Gardens in Western Australia. Local Environment 3, no. 2 (1998): 179-189.

U.S. Census Bureau. 2008 Small Area Income Poverty Estimates: Illinois, McDonough County. Accessed October 20, 2010. http://www.census.gov//did/www/saipe/county.html

U.S. Census Bureau. American Factfinder. Accessed October 20, 2010. http://factfinder.census.gov.

Chapter 13: It's Not about Food Miles: the Moral Foundations of the Local Food Movement

Brian K. Powell

Introduction

Some people are motivated to seek out local foods based on the taste; others care to strengthen local economies. However, one can also detect a significant *moral* dimension to the local foods movement. Locavores often give a look that says something like: *We are not just shopping here, we are saving the planet!* The idea is both simple and ubiquitous. We care about "food miles" because we care about our "carbon footprints" because we care about the planet Earth.[1] This argument, which I will call "the food miles argument," has found its way into popular culture and is the basis upon which many see buying local foods as an expression of moral virtue. Before addressing the merits of this argument, I want to locate the argument within a larger philosophical framework.

The Morality of Shopping in General

Shopping has not traditionally been a central topic in moral philosophy. However, in the 1970's Peter Singer started teaching us to think differently. For the last forty years, Singer has served as a sort of modern Socrates, asking us to justify the choices we often make with little thought. We go off and buy expensive coffee drinks, or handbags, or new cars, and yet we know that people are dying from lack of food, clean water, and very basic medicine. Can one morally justify spending $2000 on a handbag when that money could provide the hydration packets necessary to save the lives of dozens of children who will otherwise die? This question points to one reason why shopping is a morally significant activity:

---

[1] "Food miles," of course, refers to the number of miles food had to travel to get from the producer to the vendor. "Carbon footprint" refers to the amount of fossil fuels we are using, and thus the amount of greenhouse gases we are contributing to the atmosphere.

by choosing to spend our money in one way we are choosing not to spend it in another. This was the lesson of Singer's 1972 article "Famine, Affluence, and Morality."[2]

Another reason why shopping is a morally significant activity is that in spending our money in certain ways, and not others, we are choosing to support certain institutions and practices and not others. My coffee habit might help support the local café or it might support Starbuck's. When I shop for a dinner I plan to cook, I might help support responsible farming practices and humane treatment of animals, or I might support practices that are cruel and environmentally destructive. This was a central lesson of Singer 1975 book Animal Liberation.[3] It also connects to a third (and somewhat overlapping) reason why shopping is a morally significant activity. How we spend our money, and thus which practices and institutions we choose to support, affects the lives and wellbeing of other people, non-human animals, and the planet earth. Just think of sweatshops, factory farms, and sports utility vehicles.

If Singer is a modern day Socrates, then his message is clear: unexamined shopping is not ethically defensible. In order to discharge our ethical responsibilities, we must ask questions about the things we buy. Where did this come from? Who made it? Under what conditions, etc.? We need to know "the story" of what we buy. This can be rather overwhelming, and I think that it can be taken too far. Indeed, there is a significant dose of what Nietzsche called the "ascetic ideal"[4] in those we might call modern day "environmental saints." It has been important for me to avoid absolutism about shopping ethics; my dictum is not the Biblical "be thou perfect," but rather "be thou better."

The Morality of Food in General

The above considerations have a clear application to choices about how to shop for food. It is clear that how we feed ourselves greatly affects the lives and wellbeing of other people, non-human animals, and the planet earth. A very clear example of this is the much discussed example of factory farmed beef.

If we buy factory farmed beef, then we are supporting an institution that inflicts great suffering on non-human animals, negatively affects the wellbeing of humans, and is responsible for catastrophic environmental harms. Factory farms are places where large numbers of animals are confined close together and provided with little, if any, access to sunlight, fresh air, or room for natural movement. Factory farms are designed to produce as much meat or animal product as cheaply and quickly as possible. Thus, the wellbeing of the animals is typically given no consideration whatsoever.[5] Moreover, these facilities generate

---

[2] Peter Singer, "Famine, Affluence, and Morality," *Philosophy & Public Affairs* 1, no. 1 (Spring, 1972): 229-243.

[3] Peter Singer, *Animal Liberation* (New York: Harper, 2009).

[4] Friedrich Nietzsche, *On the Genealogy of Morality*, Trans. by Maudemarie Clark and Alan Swensen (Indianapolis: Hackett, 1998).

[5] Tom Regan, "Ethical Vegetarianism and Commercial Animal Farming," in *Today's Moral Problems*, ed. Richard Wasserstrom (New York: Macmillan, 1985), 330; Jim Mason and Peter Singer Mason, *Animal Factories* (New York:Three Rivers Press, 1990).

enormous amounts of waste, which are a significant threat to human wellbeing.[6] This manner of feeding ourselves is also very wasteful, and carbon intensive.[7] Producing beef requires an enormous amount of fossil fuels, and cow flatulence is a significant source of methane.[8] These considerations have led some to ask: which is the more significant source of greenhouse gas emissions, "the fork" or "the car key"?[9]

President Obama has called climate change "the greatest moral challenge of our time,"[10] and Singer has pointed out that climate change "will almost certainly kill more people" than things such as terrorism.[11] Climate change promises to inflict great suffering and death on humanity, and it promises to disproportionately affect the global poor.[12] Thus, there can be no doubt that climate change is a serious moral problem, and how we are producing our food is a significant cause of climate change.

The "Food Miles" Argument

This brings us back to what I've called "the food miles argument." The argument can be stated as follows. If your food comes from far away, then it must be shipped in and that will require fossil fuels and thus greenhouse emissions. We need to reduce fossil fuel use and greenhouse emissions. Thus, we need to eat local, and pay attention to our food miles.

The problems with this argument are by now fairly well known. Firstly, the number of miles that food has traveled tells one very little about the amount of energy used to get the food "to the fork." For instance, local tomatoes from a heated greenhouse will surely have a bigger carbon footprint than efficiently shipped tomatoes from a warmer climate, and driving across town to buy local broccoli is likely to be less energy efficient than picking up non-local broccoli during your weekly trip to the supermarket. Secondly, it is well established that

---

[6] Michael Fox, "The Contribution of Vegetarianism to Ecosystem Health," *Ecosystem Health* 5, no. 2 (June 1999):71; Michael Fox, "Vegetarianism and Planetary Health," *Ethics and the Environment* 5, no. 2 (2000):165; William Stephens, "Five Arguments for Vegetarianism," *Philosophy in the Contemporary World* 1, no. 4 (Winter 1994): 26-27.

[7] Fox, "The Contribution of Vegetarianism to Ecosystem Health," 70; Fox, "Vegetarianism and Planetary Health," 165; Stephens, "Five Arguments for Vegetarianism, 27.

[8] "Livestock's Long Shadow," *Food and Agriculture Organization of the United Nations* (2006), ftp://ftp.fao.org/docrep/fao/010/a0701e/a0701e00.pdf; "Eating Beef is Less Green Than Driving," *London Telegraph*, July 2007, http://www.telegraph.co.uk/news/uknews/1557846/Eating-beef-is-less-green-than-driving.html; Michael Hawthorne, "Raising a Global Stink," *Chicago Tribune*, July 31, 2008, http://articles.chicagotribune.com/2008-07-31/news/0807310206_1_carbon-dioxide-cattle-feed-methane; Stephens, "Five Arguments for Vegetarianism," 27.

[9] Claudia Deutsch, "Trying to Connect the Dinner Plate to Climate Change," *New York Times*, August 29, 2007, http://www.nytimes.com/2007/08/29/business/media/29adco.html.

[10] Henry Payne, "Crucify Them," *National Review Online*, April 30, 2012, http://www.nationalreview.com/planet-gore/297244/crucify-them-henry-payne.

[11] Peter Singer, *One World* (New Haven: Yale University Press, 2004): 1.

[12] Paul Harris, "Fairness, Responsibility, and Climate Change," *Ethics and International Affairs* (August 2006): 150.

transportation accounts for a small portion of the energy used in the process of getting ourselves fed.[13] Thirdly, focusing on food miles might distract one from more serious inefficiencies in our system, such as the inefficiency of eating higher up "the food chain." These are some of the reasons why, in the words of Historian James McWilliams, "a food-mile litmus test is an inadequate measure of our food's environmental impact."[14]

The Moral Heart of the Local Food Movement

I take myself to have established two points thus far. Firstly, shopping in general and shopping for food in particular are matters of great moral importance. Secondly, counting "food miles" is not always a rational way to approach food ethics. However, to end the discussion here would completely miss *the heart of the local foods movement.*

The moral heart of the local food movement is *not* food miles. The local food movement is about knowing the story of where your food came from. The locavores I know do not count miles so much as *cultivate relationships*. In the process, they acquire knowledge about how their food is being produced. It is no accident that one does not hear locavores gushing about the meat from the local concentrated animal feeding operation (CAFO). I would argue that the real moral power of the local foods movement is epistemological in nature. In other words, it's about knowledge. Locavores want to know how the people involved are treated, how non-human animals are treated, and how the planet is being affected. The moral heart of the local foods movement is the willingness to take responsibility for our food choices, to care about the story of our food, and to put ourselves in a position to know how our choices are affecting other people, non-human animals, and the planet earth. This makes the locavore a natural ally for other sorts of activism.

The locavore may not be an animal rights activist, but she will know where her meat comes from, and is not likely to purchase it if it is produced in a cruel manner. She may not be a labor activist, but she is likely to know, and care about, how workers are treated by those involved in producing her food. She may not be an environmentalist, but she is likely to know, and care about, how the production of her food affects the planet we all share.

References

Deutsch, Claudia. 2007. Trying to Connect the Dinner Plate to Climate Change. *New York Times*, 29 August.

Fox, Michael. 1999. The Contribution of Vegetarianism to Ecosystem Health. *Ecosystem Health* 5(2).

—. 2000. Vegetarianism and Planetary Health. *Ethics and the Environment* 5(2).

---

[13] Jim Mason and Peter Singer, *The Ethics of What We Eat* (Emmaus:Rodale, 2006):145-146. Mason and Singer discuss a study that puts the number at eleven percent.

[14] James McWilliams, *Just Food* (New York: Little, Brown, and Company, 2009): 23.

Harris, Paul. 2006. Fairness, Responsibility, and Climate Change. *Ethics and International Affairs*.
Hawthorne, Michael. 2009. Raising a Global Stink. *Chicago Tribune*, 31 July 31.
McWilliams, James. 2009. *Just Food*. New York: Little, Brown, and Company.
Nietzsche, Friedrich. 1998. *On the Genealogy of Morality*, Trans. by Maudemarie Clark and Alan Swensen. Indianapolis: Hackett.
Payne, Henry. 2012. Crucify Them. *National Review Online*. 30 April 30.
Regan, Tom. 1985. Ethical Vegetarianism and Commercial Animal Farming. In *Today's Moral Problems*, ed. Richard Wasserstrom. New York: Macmillan.
Singer, Peter. 1972. Famine, Affluence, and Morality. *Philosophy & Public Affairs* 1(1):229-243.
—. 2004. *One World*. New Haven: Yale University Press.
—. 2009. *Animal Liberation*. New York: Harper.
Singer, Peter and Jim Mason. 1990. *Animal Factories*. New York: Three Rivers Press.
—. 2006. *The Ethics of What We Eat*. Emmaus: Rodale.
Stephens, William. 1994. Five Arguments for Vegetarianism. *Philosophy in the Contemporary World* 1(4).
Livestock's Long Shadow. 2006. Food and Agriculture Organization of the United Nations.
Eating Beef is Less Green than Driving.2007. *London Telegraph*, July.

Chapter 14: What People are for: Agrarianism and Religion in the Heartland

Betsy Perabo

Introduction

Native American religions are sometimes admired for their integration of spirituality and place. Their practices and beliefs are moored to the land, in an apparent contrast to faiths that strive to be timeless and placeless. Yet in certain ways all religions adapt to their environment, especially in rural areas, where they are far from their origins and from official centers of authority. In rural America, these adaptations may be most striking among religious minorities, where a half dozen Buddhists gather to chant in a converted dining room of a member's house, or where Muslims enlist an Amish farmer to provide freshly slaughtered meat for a ritual, but Christians may also be said to adapt their religion to their environment. Rural ministers from across the Christian denominations emphasize their congregants' profound sense of place, and especially their connection to their land and their farms.[1]

Sociologists and other observers suggest these adaptations of Christianity to place have been occurring for a long time. Mary Jo Neitz speaks of the symbolic status of rural churches, including those she has studied in Missouri. The idea that the sacred resides in nature is at the core of this status. The belief – or myth - is that "Due to their spatial location, close to nature, and perhaps far from urban areas, rural churches can be seen as carriers of something purer and closer to God."[2] Residents come to see their land as "God's country" in some way that other places are not, and this affects their religious lives. In her sketch of spiritual life on the Great Plains, Kathleen Norris suggests that the isolation and openness make the locale like a monastery and its residents voluntary or involuntary

[1] See, e.g., Miriam Brown, ed., *Sustaining Heart in the Heartland: Exploring Rural Spirituality* (New York/Mahwah: Paulist Press, 2005), and Shannon Jung et al., *Rural Ministry: The Shape of the Renewal to Come* (Nashville: Abingdon, 1998).
[2] Mary Jo Neitz, "Reflections on Religion and Place: Rural Churches and American Religion," *Journal for the Scientific Study of Religion* 44(3) (2005), 243.

ascetics. The people in one rural church she describes "have become one with their place," and what makes their church distinctive is that almost all of its members are "totally dependent on the land."[3]

This sense of the value of the rural, and of the "heartland" – a term that can mean either the physical center of a region, or its most important part – may also be linked to the land's ability to provide food.

One of the heroes of Michael Pollan's *The Omnivore's Dilemma* is Virginia farmer Joel Salatin, who is both a committed evangelical Christian and a committed locavore; Salatin speaks of farming as his "ministry" and customers as "parishioners." [4] His commitment to locavorism is so strong that he refuses to ship his farm's products to other regions. When Pollen notes the difficulties this principle could create for the residents of big cities, and asks how New York City could fit in his vision of sustainability, Salatin replies: "Why do we have to have a New York City? What good is it?"[5]

For some people, then, rural areas are ideal sites with respect to both religion and farming. But what is the connection between these two things? Attempts to answer this question have generated a veritable cottage industry surrounding local food and sustainability among the dominant, non-Native religious traditions in the United States and within the academic field of religious studies. In the 1970s, ecological concerns were most strongly associated with liberal Christians, but the movement has become more diverse, gaining support among some evangelicals and drawing attention from academics seeking to situate it in different biblical and theological frameworks. Religious language has also seeped into discussions of sustainability by writers whose work otherwise seems entirely secular.

In the broadest sense, the critical question for understanding the relationship between religion and farming is: "what are people for?" This is a question Wendell Berry first posed in an essay criticizing the United States for ignoring the human impact of its agricultural policy, but it is also what scholars of religion call an "anthropological" question, anthropology here meaning the study of the true nature of human beings. The belief that people have a particular purpose dates back thousands of years; Aristotle, for example, suggested that the all things have a *telos*, or end, that they are best suited to fulfill. Berry is a strong advocate of "agrarianism," which according to one definition is "a social and political philosophy that describes the cultivation of plants and the care of animals, or farming, as a means to a fuller and happier life".[6] According to this definition, farming makes it possible for us to reach or achieve our *telos* or end. There are many definitions of agrarianism, and they share this idea of the centrality of the land, of farming, and of humanity's place in the natural world. However, as I will

---

[3] Kathleen Norris, *Dakota: A Spiritual Geography* (Boston: Houghton Mifflin, 1993), 169.

[4] Michael Pollan, *The Omnivore's Dilemma: A Natural History of Four Meals* (New York: Penguin, 2006), 240-241.

[5] Ibid., 245. Pollan sees this "deep antipathy to cities that has so often shadowed rural populism" as a possible "dark side to "Joel's vision of the postindustrial food chain."

[6] Sherry Williamson, "Norman Wirzba on Taking Care of Creation," *Divinity* 8:2 (Winter 2009), http://www.divinity.duke.edu/publications/2009.01/features/feature5/index.htm.

discuss below, some definitions are explicitly religious, others are implicitly religious, and others do not mention God or religion at all.

The basic agrarian argument of the non-native religious traditions in the U.S. – and in this essay I will focus on Christianity – is that if we wish to be close to God, and if we wish to fulfill our God-appointed *telos*, we must attend to agrarian concerns. It is possible that this is best done, or easiest to do, in the rural areas of the heartland where farming is a central occupation; it could be argued, that here, locavorism may be a means of making a virtue of necessity. This essay will review the literature that addresses religion's role in the local food and sustainability movements, and situate it in rural environments.

The influence of Kentucky farmer and ecologist Wendell Berry on these issues has been deep and wide-ranging, and the essay begins with his analysis of the spiritual roots and justifications of agrarianism. I then look at the work of two advocates of local food from different parts of the country and of the Christian spectrum: Presbyterian minister Craig Goodwin, who described his family's efforts to eat locally in *The Year of Plenty*, and Joel Salatin, a farmer featured in *The Omnivore's Dilemma*, who received Bob Jones University's Alumnus of the Year Award in 2009. Even when not discussing Salatin, Pollan frequently uses religious language in *The Omnivore's Dilemma,* and I will also consider how that language functions in the context of a secular discussion of these issues. Rachel Wheeler notes that many supporters of the local food movement, whether or not they are religious in a traditional sense, use language in this way, in what she calls "locavangelism." I then describe the work of theologian Norman Wirzba, who provides a theological framework integrating agrarianism with traditional Christian concepts such as eucharist, creation, and idolatry, and biblical scholar Ellen Davis, who suggests that the Hebrew scriptures contain an agrarian perspective important for modern Americans.

Wendell Berry and the Development of "Agrarianism"

Wendell Berry's work on agrarianism is so extensive and influential that it is difficult to summarize. He has been writing on ecological issues since the 1960s, and in 2002 Norman Wirzba, collected his essays on these topics in *The Art of the Commonplace: The Agrarian Essays of Wendell Berry*. Given his holistic perspective, Berry might be loath to try to separate out "religion" from the other components of his thought, but the collection does contain a section that is called "Agrarian Religion." I will focus on the essays in this section, as well as note a few other places where Berry writes and speaks about religious and spiritual concerns.

In many of his essays, Berry refers to but does not define "agrarianism," and, as noted in the introduction, the term has come to have several definitions. However, in his essay "The Whole Horse" (2002) he sketches out a complex description of it that stretches out over four pages. He begins by noting that agrarianism is not primarily an "idea" but "a practice, a set of attitudes, a loyalty, and a passion," and says though he was "raised by agrarians," he did not hear the term defined until he was in college [7] He begins by saying that as opposed to industrialism, which is "a way of thought based on monetary capital and technology, agrarianism is a way of thought based on land."[8]. Centered on the household, the agrarian economy focuses on local realities rather than abstract principles. It depends on nature, he says: "In the written records of agrarianism, there is a continually recurring affirmation of nature as the final judge, law-giver, and pattern-maker of and for the human use of the earth."[9] It is not until the fourth page that Berry mentions God as part of agrarianism, though his earlier references to "nature" and "Creation" may signal God's presence or activity. He says here, "The agrarian mind begins with the love for the fields and ramifies in good farming, good cooking, good eating, and gratitude to God."[10] In the works of others following in Berry's tradition, it is not clear whether God or religion is a necessary part of agrarianism. Davis, in a work introduced by Berry, says it "is a way of thinking and ordering life in community that is based on the health of the land and of living creatures."[11] Wirzba is more explicit about agrarian's religious foundations, stating in his introduction to Berry's work that agrarianism "promises a path towards wholeness with the earth, with each other, and with God, a path founded upon an insight into our proper place within the wider universe."[12]

Whether or not God is necessarily part of agrarianism, Berry himself is clear elsewhere, and from a very early stage, that attention to farming and food is a means of religious practice. In "The Use of Energy" (1977), Berry suggests that farming is both a "practical art" and "a practical religion, a practice of religion, a rite."[13] "By farming," he says, "we enact our fundamental connection with energy and matter, light and darkness. In the cycles of farming, which carry the elemental energy again and again through the season and bodies of living things, we recognize the only infinitude within the reach of the imagination."[14] As we participate in this cycle, "we align ourselves with the universal law that brought the cycles into being and that will survive them."[15] He notes the root "cult" in both "agriculture" and "cultivation": "The ideas of tillage and worship are thus joined in *culture*... To live, to survive on the earth, to care for the soil, and to worship, all are bound at the root to the idea of a cycle."[16] If we remain in direct contact with these basic sources of life, we will develop an awareness of our limits. He holds up the Amish as an example of the effectiveness of abstaining from the use of technological power, and says the fact that they are flourishing

---

[7] Wendell Berry, "The Whole Horse," in *The Art of the Commonplace: The Agrarian Essays of Wendell Berry*, ed. Norman Wirzba (Berkeley: Counterpoint, 2002), 238.
[8] Ibid., 239.
[9] Ibid., 240.
[10] Ibid., 241.
[11] Ellen Davis *Scripture, Culture, and Agriculture: An Agrarian Reading of the Bible* (Cambridge: Cambridge University Press, 2009), 1.
[12] Norman Wirzba, introduction to *The Art of the Commonplace: The Agrarian Essays of Wendall Berry*, viii.
[13] Berry, "The Use of Energy," in *The Art of the Commonplace*, 285.
[14] Ibid.
[15] Ibid.
[16] Ibid.

indicates that "we can make ourselves whole only by accepting our partiality, by living within our limits, by being human – not by trying to be gods. By restraint they make themselves whole."[17]

Berry also identifies biblical themes that directly support these ideas. In "The Gift of Good Land," (1979) Berry suggests that the biblical accounts of creation, and especially of the responsibilities the Israelites have for the Promised Land, support the idea that God has given human beings the land as a gift. When they receive the land, the Israelites are warned against *hubris* and reminded of their dependence on God. They are not gods, and if they act like gods they will be punished. They must use the land properly and bear in mind that ultimately the land belongs to God. While not denying Christian indifference to and complicity in the abuse of nature, Berry argues that these scriptures constitute a critical part of the Jewish and Christian heritage, and suggests that they give Jews and Christians a form of the Buddhist doctrine of "right livelihood"; we are at our best when we find occupations that cohere with God's intentions for humanity.

For Berry, then, agrarianism is compatible with Christianity. More than that, though, he suggests that agrarian questions are in essence religious questions. If farming – which involves the use of our bodies to bring food from the land - is a practical religion, and the land is a gift to us, both body and land must be understood in the broader context of God's creation and ongoing creative and sustaining acts. This is linked to the anthropological question mentioned earlier: what are people for? He begins the essay "The Body and the Earth" (1977) by saying:

The question of human limits, of the proper definition and place of human beings within the order of Creation, finally rests upon our attitude toward our biological existence, the life of the body in this world. What value and respect do we give to our bodies? What uses do we have for them? What relation do we see, if any, between body and mind, or body and soul? What connections or responsibilities do we maintain between our bodies and the earth? These are religious questions, obviously, for our bodies are part of the Creation, and they involve us in all the issues of mystery.[18]

Humans beings understand themselves as small within creation, and "Such humility is the consequence of an accurate insight, ecological in its bearing, not a pious deference to 'spiritual' value."[19] That is, the "religious" insight that we are part of creation is confirmed by our experience of the understanding of our place in the world that we gain for agriculture. Our health, then, combines the physical, mental, and spiritual. The word health, he says, "belongs to a family of words, a listing of which will suggest how far the consideration of health must carry us: *heal, whole, wholesome, hale, hallow, holy*."[20] The health of the body is the health of the soul, and we should not marginalize either the body or the earth; physical work is not something to be avoided but is important for body and soul.[21] Berry stresses the importance of the local in this essay, saying, "There can be no

---

[17] Ibid., 292.
[18] Berry, "The Body and the Earth," in *The Art of the Commonplace,* 93.
[19] Ibid., 94.
[20] Ibid., 98-99.
[21] Ibid., 102

such thing as a 'global village.' No matter how much one may love the world as a whole, one can live fully in it only by living responsibly in some small part of it."[22]

Berry's holistic perspective is once again raised in a 1983 essay, "Two Economies," where he discusses how the most comprehensive way to describe or conceive of the economy would be to think of it, or understand it as, the Kingdom of God. We cannot conceive the economy as simply an "industrial economy" – this is not comprehensive enough - and to critique it we need the concept of an economy that does not leave anything out is useful. In the Kingdom of God, for example, "the fall of every sparrow is a significant event. We are in it whether we know it or not and whether we wish to be or not."[23] Everything is joined together, but we do not know either all it contains or how it is ordered. He suggests that this is not an exclusively biblical concept, seeing it in ancient Greece, and in the eastern concept of the Tao, and both to reflect that and to create a more "culturally neutral term" he refers to it as the Great Economy.[24] (Feminist theologian Letty Russell prefers to use the term "household of God" rather than "kingdom of God," and this fits in well with Berry's conception.)

Berry does a more comprehensive analysis of Christianity's relationship to agrarianism in his 1992 essay, "Christianity and the Survival of Creation." This essay, delivered to the Southern Baptist Theological Seminary in Kentucky, draws together a number of the themes mentioned above. He continues to insist that the Bible does not support environmental destruction, and that those who have Christianity as their "native religion" must keep it alive in a renewed way that helps the earth flourish. God, after all, loves the world; John 3:16 ("For God so loved the world that he gave it his only son") means that Christ came precisely because of God's love for the world. As suggested by the idea of "right livelihood" mentioned earlier, human work should honor the work of God.[25] God as craftsman and creator should be our model for crafting and creating.[26] Reiterating the connection between religion, the land, and the production of food, he notes that the outdoors is the site of so much that is miraculous:

Whoever really has considered the lilies of the field or the birds of the air and pondered the improbability of their existence in this warm world within the cold and empty stellar distances will hardly balk at the turning of water into wine – which was, after all, a very small miracle. We forget the greater and still continuing miracle by which water (with soil and sunlight) is turned into grapes.[27]

Berry's themes have been consistent throughout the years, and if anything Berry has become more focused on local issues. In a 2004 interview with the Christian magazine *Sojourners*, Berry suggests that local food is at the core of his agrarian concerns. He says "first lesson to learn about agriculture" is that "It needs a sound subsistence basis. People need to feed themselves, next they need

---

[22] Ibid., 118.
[23] Berry, "Two Economies," in *The Art of the Commonplace,* 220.
[24] Ibid., 221.
[25] Berry, "Christianity and the Survival of Creation," in *The Art of the Commonplace,* 312.
[26] Ibid., 317.
[27] Ibid., 311.

to feed their own communities. That's what we're working for now."[28] In a 2009 introduction to Ellen Davis' *Scripture, Culture, and Agriculture*, he writes , "From an ecological and agrarian point of view, the most urgent problem of agriculture, as of the human economy as a whole, is that of local adaptation – that is to say, of making a beneficent and conserving fit between work and place." [29]

The Fruits of Berry's Locavorism: Christian Consumers and Producers

Berry's work, rooted in his home in Kentucky, has helped to give rise to the locavore movement throughout the United States. Some people have supported the movement from a more explicitly Christian perspective than Berry has. Christian magazines such as *Sojourners* and *The Christian Century* publish articles on local food and sustainability, and there are many examples of groups throughout the country that are engaged in bringing religion and food issues together. In the Midwest, for example, the group Earth Keepers, founded by two Lutheran churches in the Quad Cities area, focuses on agrarian concerns,[30] and Western Theological Seminary in Holland, Michigan, held a conference in 2010 called "The Earth is the Lord's" to help Christians care better for creation.). The participants in the Rural Spirituality Team of the Churches' Center for Land and People in Wisconsin, note that the development of more "land-based groups" and the convening of conferences with titles such as Reclaiming the Sacred in Farming and Food,' "Holy Soil, Common Ground,' 'Spirituality and the Land.'"[31]

There are many other examples that might be given. Here, however, I will concentrate on only two of Christians who are active and very visible in the movement: Craig Goodwin, who is a Presbyterian minister in Washington state; and Joel Salatin, who is a self-identified ""Christian-conservative-libertarian-environmentalist-lunatic farmer" and operates Polyface Farms in Virginia.[32]

Goodwin is the author of *Year of Plenty: One Suburban Family, Four Rules, and 365 Days of Homegrown Adventure in Pursuit of Christian Living*, which was published in 2011. In an appendix to his book, titled "A Brief Explanation of My Obsession with Wendell Berry," Goodwin describes Berry as a "tour guide for our family as we navigated through the issues of food, land, farming, and faith."[33] At the end of 2007, Goodwin and his wife decided that for a year, they and their children would limit their consumption by only using goods that were

---

[28] Rose Marie Berger, "Wendell Berry interview complete text," *Sojourners* 33:7 (July 2004), http://www.sojo.net/index.cfm?action=magazine.article&issue=soj0407&article=040710x.

[29] Wendell Berry, forward to *Scripture, Culture, and Agriculture: An Agrarian Reading of the Bible* by Ellen F. Davis (Cambridge: Cambridge University Press, 2009), xii.

[30] "Faith in Action: Earth Keepers answers the call to be stewards of creation." *Radish: Healthy Living from the Ground Up*, June 2012, 28.

[31] Brown, *Sustaining Heart in the Heartland*, 27.

[32] Pollan, *Omnivore's Dilemma*, 125.

[33] Craig Goodwin, *Year of Plenty: One Suburban Family, Four Rules, and 365 Days of Homegrown Adventure in Pursuit of Christian Living* (Minneapolis: Sparkhouse Press, 2011), 206.

locally-produced, used, homegrown, homemade, or (both as a means to participate in the world economy and as a concession to their desire to have coffee) from Thailand. *Year of Plenty* is not strictly about an experiment in locavorism, then, but it frequently focuses on this issue.

Goodwin makes it clear from the outset that the experiment is not, on the one hand, simply a Christian adaptation of an environmental "trend" prominent in the culture – "Our intent is not to take a cultural narrative and shellac a Jesus fish on it," he says.[34] He believes that there is biblical support for changing one's lifestyle in this way. By the same token, he stresses that in embracing these principles, he is not simply trying to lure in more congregants to the church. He writes, "my interest in food and consumption is not some bait-and-switch effort to slip Jesus into people's lives, as if local food were some carrot on a stick to lead people along into the holy. The whole point is that I am learning to pay attention to real carrots, preferably local and organic, and see them as holy in some way."[35]

In describing this experiment, Goodwin highlights many of the same issues Berry raises. He believes that developing relationships with local producers humanizes the process of consumption, creating more opportunities to know and love one's neighbors. Grace before meals becomes "an inventory not just of the food but of the people whose work had gone into producing it."[36] Early in the year, before a meal supplied by local producers who they had made an effort to meet, Goodwin says, "Every food item on our plates was connected to a person we knew, a person who was sorting out a life of hopes and dreams and struggles. These were our neighbors, and on that night, through our prayers, we were able to love them as never before."[37] Later in the year, they help to coordinate a local farmer's market in the parking lot of the church, and again focus on the importance of local food for creating community.[38]

For Goodwin, there are two major connections between Christianity and the local foods movement. The first, described above, is communal. He suggests that the farmer's market is a good metaphor for the kingdom of God, and asks: "what if our whole life was like a farmer's market, where we knew and care for the people who produce the things we consume?"[39]

The second connection has more to do with individual development, though it has impact on the community as well. Goodwin sees his reliance on local foods as means of *askesis,* "an intentional exercise of entering into limitations with a goal of growth and maturity" (the root of the word ascetisim).[40] Gardening, for example, leads people to cultivate virtues such as humility and gives them a deeper sense of the miraculous nature of life;[41] and the greater attentiveness to seasons, to weather, and to the land around us gives us each a better sense of our place in the universe. As Berry notes, the person who grows a garden "will no

---

[34] Ibid., 14.
[35] Ibid., 138.
[36] Ibid., 35.
[37] Ibid.
[38] Ibid., 76.
[39] Ibid., 83-84.
[40] Ibid., 103.
[41] Ibid., 119.

longer look upon rain as a traffic impediment, or upon the sun as a holiday decoration."[42] While it is possible to read the curse of Adam and Eve in Genesis as simply punitive, Goodwin argues that our involvement in growing food aids in forming our characters.[43] He states that as he and his family began to convert their lawn into a garden, he had a sense that "we were somehow more faithfully occupying our place in the order of creation."[44] Their garden "was becoming a place of formation, both a training ground for the journey and a living metaphor of pilgrimage to the Creator: Father, Son, and Holy Spirit."[45]

This knowledge that comes from *askesis* enables those who engage in these kinds of efforts to enter the life of consuming – which is inescapable – as one enters the kingdom of God, and to think about different ways in which the kingdom of God manifests itself on earth. He says, "And in this entering, we become poetic miners of God's kingdom, listening to the emerging story, and hopefully giving it a voice; the kingdom of God is like a farmers' market, the kingdom of God is like a CSA box, the kingdom of God is like a farm with happy chickens, the kingdom of God is like.... That's how Jesus entered and told and enacted the story. That is my understanding of how we also are called to enter the world." [46]

In concluding, Goodwin reiterates Berry's statement that "Eating is an agricultural act," but expands on this idea, stating, "eating is a theological act, with deep connections not only to land but to the Creator of the land. More than that, buying food, or anything else, is a theological act."[47] Christians bring a unique perspective to this act, he says: "this faith in the God who is saving the world is the church's greatest gift to an environmental movement that struggles to make sense of small actions in the face of global crises."[48]

Joel Salatin approaches the local foods movement from the perspective of a producer rather than a consumer, but in many respects his concerns are the same as Goodwin's. In *The Omnivore's Dilemma,* Michael Pollan describes a dinner he eats with the Salatin family at Polyface Farms as "probably the all-time most local meal I'd ever eaten," with almost everything grown at the farm itself.[49] Like Goodwin, Joel Salatin is deeply aware of the sources of the food he eats and concerned that others are as well, and the grace he says at the beginning of his meal resembles the one Goodwin talks about above. Pollan notes, "Joel began the meal by closing his eyes and saying a rambling and strikingly non-generic version of grace, offering a fairly detailed summary of the day's doings to a Lord who, to judge by Joel's tone of easy familiarity, was present and keenly interested."[50] In Salatin's case, however, he is focusing on his own activities rather than those of other providers.

---

[42] Ibid.
[43] Ibid., 88.
[44] Ibid., 120.
[45] Ibid.
[46] Ibid., 134
[47] Ibid., 195.
[48] Ibid., 196.
[49] Pollan, *Omnivore's Dilemma*, 203.
[50] Ibid., 203.

Salatin is widely known for his concerns about local food and his religious beliefs. The Polyface Farms web site states, "We are in the redemption business: healing the land, healing the food, healing the economy, and healing the culture."[51]. According to Fred Walters, the editor of AcresUSA, Salatin is "sort of the high prophet of sustainable family farming."[52] Salatin is a graduate of Bob Jones University, an evangelical Christian university which states that it is "committed to defending and teaching the inerrancy of Scripture and consistently applying biblical truth to every arena of life."[53] In the 1970s, students there were told to avoid "the food cult," which was seen as a countercultural leftist movement. BJU has apparently changed its tune about "the food cult," as Salatin received its 2009 alumnus of the year award.

Salatin now wants the local-foods movement to be a "big tent," and is pleased that his consumers come from across the political and religious spectrum. However, he does not see evangelicals embracing the local foods movement as much as he would like. He "laments that he cannot convince more of his fellow churchgoers not to "stop for happy meals on the way home from the pro-life rally."[54] His stridently anti-government intervention attitudes make him popular among some conservatives and libertarians – he has not attended to get "organic" certification from the federal government, and says he wants to create "an NRA for local food" by starting a legal defense fund for farmers who may be harmed by government regulations. Pollan notes that he opposes efforts to legislate good behavior with regard to food – ""we ask for too much salvation by legislation," he says[55] – and instead calls for individual action. Salatin compares the food movement to the Protestant Reformation: "It's a little like Luther nailing his ninety-five theses up at Wittenberg. Back then it was the printing press that allowed the Protestants to break off and form their own communities; now it's the Internet, splintering us into tribes that want to go their own way."[56] The industrial

---

[51] Polyface Farms, http://www.polyfacefarms.com/.

[52] Todd S. Purdum, "High Priest of the Pasture," *New York Times,* May 1, 2005, http://query.nytimes.com/gst/fullpage.html?res=9D0CE7DF173EF932A35756C0A9639C8B63. AcresUSA, while not a religious organization, has raised religious concerns about such issues as the Farm Bureau's 2008 requirement to register all cattle through the National Animal Identification System. Charles Walters opens a December 2005 article, titled "The Mark of the Beast," by quoting Revelation 13:15-18, and suggests that if computer chips are planted in every animal, "In a generation or two the chip will be planted in every human being.". Another article notes that Amish farmers "had their Bibles in tow" when attending a Michigan meeting about the regulation, and states, "The First Amendment is savaged by the NAIS. Many Christians are forbidden to accept the mandatory mark because it violates scripture and therefore violates freedom of religion" (Acres USA 36:12 (December 2006)). Though Salatin does not mention the biblical concern anywhere I have been able to locate, he does express a deep disdain for many forms of government regulation.

[53] Bob Jones University, "Why Choose BJU?" http://www.bju.edu/why-choose-bju/.

[54] Lewis McCrary, "Joel Salatin practices ethical animal husbandry— no thanks to the feds," *The American Conservative*, November 2009, http://amconmag.com/article/2009/nov/01/00023/.

[55] Pollan, *Omnivore's Dilemma,* 260.

[56] Ibid., 261.

food industry will be like the Catholic priests who witnessed the exodus of their congregations.

Though Salatin's rhetoric is more revolutionary than Goodwin's, the theological analysis of food is somewhat similar. On the farm, Salatin grows a variety of crops and raises several different animals, and does what he calls "stacking" to make the system work more efficiently. For example, Salatin does not clean out the cow manure from the barn over the winter; he just adds woodchips, straw, and corn. Having this layer of "compost" as bedding keeps the barn warm. In the spring, when the cows go out to pasture, Salatin sends the pigs in and, because they want the corn, they go through the bedding, which results in aerating it to produce excellent compost (he calls this "pigaeration"). Pollan, witnessing this process, says "these were the happiest pigs I'd ever seen."[57]

There is a reason for this, Salatin says: the holistic model he follows on his farm is "the model God used in building nature."[58] He suggests that he is letting animals behave as they were meant to behave. On his farm he honors the "innate distinctive desires of a chicken" and the "natural predilections of the pig."[59] For Salatin the question is not only "what are people for?" but "what are pigs for?" and "what are chickens for?": everything has its place in the natural order, and farmers who recognize this will flourish. Salatin refuses to ship his meat, poultry, and produce, insisting that people who live outside the region support their local farmers so that this kind of farm can exist in other areas as well. Eating local, then, is a way of supporting these holistic farms and, in turn, the natural order of creation.

Pollan also uses the language of "the natural" as a consumer of the chicken; in describing a meal at the farm, he says that the chicken "smelled and tasted exactly like chicken" and "chimed with that capitalized idea of Chicken we hold in our heads."[60] Pollan's toast before the meal of chicken recognizes Salatin and his family as well as the chickens, and he calls this toast "[m]y secular version of grace, I suppose, acknowledging the various material and karmic debts incurred by this meal."[61]

Locavangelism and the Use of Religious Language

Pollan does not discuss his own religion in *The Omnivore's Dilemma*, but he often uses religious language when discussing his food choices. In an article called "The Holy Church of Food," which discusses another work of Pollan's, Laura Shapiro says Pollan is "a believer... and though he writes for the rest of us, he can't quite bring himself to take us seriously unless we can prove we've been born again."[62] She points to his extreme suggestions – buying a hog-sized freezer or going out to forage for wild greens –but says that "these bouts of the surreal

[57] Ibid., 218.
[58] Ibid., 215.
[59] Ibid., 215, 219.
[60] Ibid., 271.
[61] Ibid., 270.
[62] Laura Shapiro, "The Holy Church of Food," *Slate*, December 31, 2007.

don't reflect his politics, they reflect his religion—the holy, catholic, and apostolic church of food, where only martyrs and lost souls have to shop at Safeway."[63]

Though it may not be fair to call this Pollan's religion, certainly *The Omnivore's Dilemma* includes many examples of religious language. Sometimes it is used in relation to food in general: in his introductory chapter he suggests that we have a "Manichean view of food" and suggests that we may need to pay the "full karmic price" of a meal.[64] Although he does refer to the biblical notion that "all flesh is grass" in the chapter on organic farming, the discussions of eating animals and of hunting generate even more theological questions and include more theological language. [65] He notes that Salatin believes it is acceptable to kill chickens because they are not created in God's image and don't have souls; Pollan himself, in contrast, says cleaning the pig that he has killed is troubling because he "lacks the religious certainty that humans have souls and animals don't."[66] Despite this, the lack of any religious certainly, however, he sees the pig he has killed as "a gift – from whom or what I couldn't say."[67] This experience raises anthropological questions for him: he notes that the virtue of hunting is that "It puts large questions about who we and the animals are, and the nature of our respective deaths, squarely before the hunter."[68] Gathering, too, seems to be a theologically significant activity, in that he feels gratitude akin to that he feels upon killing a pig. It is all the more so because the work you do is not what "earns" you the food you've found. In a garden, you feel that your work has paid off, but the yield of gathering, he says," felt more like something for nothing, a wondrous and unaccountable gift." [69]

At the end of the book, when he describes his perfect meal, the one that includes only products of his hunting, gathering, gardening and other activities, he begins to say a kind of grace before the meal but stops himself. He states, "I had actually wanted to say something more, to express a wider gratitude for the meal we were about to eat, but I was afraid that to offer works of thanks for the pig and the mushrooms and the forest and the garden would come off sounding corny and, worse, might ruin some appetites. The words I was reaching for, of course, were the words of grace."[70] Nevertheless, the meal itself felt like "a ceremony"; it had become, he said, "a thanksgiving or a secular seder, for every item on our plates pointed somewhere else, almost sacramentally, telling a little story about nature or community or even the sacred, for mystery was very often the theme. Such storied food can feed us both body and soul, the threads of narrative knitting us together as a group, and knitting the group into the larger fabric of the given world."[71] He concludes by saying that the perfect meal "is one that's been fully paid for" and that despite the difficult, it might be worthy having

---

[63] Ibid.

[64] Ibid., 4, 9. Manicheanism is a theological doctrine that suggests, among other things, the division of body and soul.

[65] Ibid., 123.

[66] Ibid., 331, 358.

[67] Ibid., 355.

[68] Ibid., 358.

[69] Ibid., 390.

[70] Ibid., 406.

[71] Ibid., 408.

sometimes to "remind us of the true costs of the things we take for granted."[72] That is, we might do it as a ritual, "as slow and storied as the Passover seder."[73] He concludes, "we eat by the grace of nature, not industry, and what we're eating is never anything more or less than the body of the world."[74]

For Pollan, then, the ideal meal of a locavore is a kind of ritual, and the experiences one gains in gathering or growing the food to supply it are in some respects religious. This use of religious language is not unusual among locavores, according to Rachel Wheeler, a historian of American religion. In her analysis of what she calls "locavangelism," Wheeler mentions a number of opponents of the industrial food system who identify themselves as Christian, and others who do not, but suggests that they share a rhetorical style, that of the evangelist.[75] Like evangelists, they tell stories that resemble conversion narratives. Noting that "the movement has gained substantial momentum among two groups who rarely care to be associated with each other: evangelical Christians and secular liberals," she suggests that the two share concerns for realizing realize their beliefs through actions, or "practice." These practices shape their development as people, and sometimes as Christians, in the way that Berry describes in his work.

Wheeler describes a lecture she attended by Howard Lyman, a rancher who grew up as a supporter of the industrial food system but who now calls himself the "Mad Vegetarian Cowboy.":

> I couldn't help but feel I had just attended a religious revival. Lyman's talk had all the hallmarks of a revivalist sermon, minus any mention of God or Jesus. He had told of the sinful ways in his youth, his arrogance and his disregard for the wisdom of tradition. He recounted the crisis sparked by illness, a miraculous cure, and the epiphany that allowed him to see the error of his former ways. He then chronicled his path of righteousness. The lecture ended with what felt like an altar call, as Lyman exhorted listeners to renounce the sinful ways of the world and follow the narrow path of righteous eating.[76]

Wheeler suggests that "aspects of the contemporary environmental movement draw heavily – and largely unconsciously—from the deeply rooted American

---

[72] Ibid., 409.
[73] Ibid., 411.
[74] Ibid.
[75] Rachel Wheeler, Eating as Spiritual Practice: Locavangelism and the Evangelical Tradition in America: Part I," Religion in American History, November 25, 2011, http://usreligion.blogspot.com/2011/11/eating-as-spiritual-practice.html. Wheeler mentions a variety of secular locavangelist works, including Collin Beaven's, No Impact Man and Barbara Kingsolver's, Animal Vegetable Miracle, as well as Pollan, who she calls the "bishop of locavangelism, and says "there is an equally voluminous output by Christian publishers, though it has gone largely un-noticed by the mainstream media." She notes Salatin and Goodwin's works as well as Ragan Sutterfield's *Farming as a Spiritual Discipline* and Don Colbert's, *What Would Jesus Eat?*
[76] Ibid.

tradition of evangelicalism."[77] She gives numerous examples of the evangelical attitude taken by those who support radical changes in the food supply system. Just as Salatin regularly incorporates religious metaphors into his description of agricultural processes - cow manure transformed into compost is "transubstantiation," for example - activists identify their own transformation into opponents of big, industrial food systems as in some respects holy.[78] They encourage others to follow their path and be "born again" as people who eat responsibly and with concern for the world around them.

Agrarianism and Contemporary Theology

Writers such as Wendell Berry and others in the various foods movements have affected the academic discipline of religious studies, with a number of contemporary thinkers studying the ways in which food consumption and production relates to religion. Two recent works, Ellen Davis' *Scripture, Culture, and Agriculture*, and Norman Wirzba's *Food and Faith*: *A Theology of Eating* note their indebtedness to Berry's work. Davis, a biblical scholar, emphasizes the ways in which the Hebrew Bible, sacred to both Jews and Christians, is connected to agrarianism, and Wirzba, a theologian and a professor of "Theology, Ecology and Rural Life" at Duke Divinity School, elaborates on many of Berry's ideas.

Wirzba develops several themes in Berry's work. He supports the idea that Christianity tells us the world is God's, not our own; and as God's creation, the world is good. Building on Berry's idea that human work should honor the work of God, he says that God is a gardener and humans should follow God's model; "we learn where and who we are by becoming tillers and keepers of God's edible garden."[79] He also stresses the importance of the communal nature of life in God's creation. This is true not just of farming but of all eating; "When it is done in the name of God, he says, "eating is the earthly realization of God's eternal communion-building love."[80]

Wirzba is not saying that we should behave properly with regard to food and farming because God has commanded us to, or wants us to. The most important thing about eating, gardening, and farming more responsibly is that it will put us more deeply in touch with God. This is connected with Berry's ideas about "what people are for." Gardens, and by extension farms, are "the practical sites in terms of which people begin to see, smell, hear, touch and taste the breadth and depth of human membership and responsibility."[81] Though one of the central roles of eating is to remind us of our dependence on God – the fact that we like all creatures must eat is both "humbling" and "terrifying" – being mindful of this fact enriches our lives.

---

[77] Rachel Wheeler, "Eating as Spiritual Practice: Locavangelism and the Evangelical Tradition in America: Part II." Religion in American History, November 28, 2011, http://usreligion.blogspot.com/2011/11/eating-as-spiritual-practice_28.html.
[78] Salatin's reference appears in Pollan, *Omnivore's Dilemma*, 218.
[79] Wirzba, *Food and Faith,* xiii.
[80] Ibid., xiv.
[81] Ibid., xvi, 36n.

Our ability to control food production tempts us to be arrogant, but we learn from work in farms and gardens that we are not gods.[82] When the food system is industrialized, this temptation is even greater; people can eat bread without knowing where its ingredients were grown or how it was made. This "deep severing between people and their food-providing social and ecological contexts" has been harmful, both physically and theologically.[83] To correct this, Wirzba recommends that eating be considered as a "spiritual exercise," something that develops habits that enable people "to live a more ordered, measured, reflective, free, attentive, available, and responsible life."[84]

Wirzba's view of the religious significance of "local food" (though he does not call it that) is more abstractly stated than Berry's. He suggests that human beings are defined, in a sense, by the place in which they live. He says, "we cannot ever be *without* a place and be well. To be alive is to be the beneficiary of the places that literally and figuratively feed us."[85] First and foremost, the biblical story in Genesis tells us that "humanity (*adam)* is what it is, *is* at all, because of its relation to soil (*adamah*) and the life God makes possible through it."[86] But more specifically, we feel our relationship to a particular place by growing and eating the foods that are unique to it. This connection not only helps us fit into that place, but helps us see our part in creation as a whole.

Though Wirzba does not link this view of place into the locavore movement, the connections are clear. If, as he suggests, we are this deeply linked to the places where we live, relying on foods produced locally or even foods we produce ourselves would seem to be something like a theological imperative.

This deep linkage between people and land is a central theme in the Bible, Davis suggests. She argues that "Beginning with the first chapter of Genesis, there is no extensive exploration of the relationship between God and humanity that does not factor the land and its fertility into that relationship."[87] Americans who assume that they can get entirely out of the agriculture business – as has been suggested by some modern economists – need to realize that the biblical writers have an "agrarian mindset." The biblical writers emphasize that the land is an integral part of human life. We "work" or "till" the land, but in a sense we also serve it; we "keep" the garden but also observe it and learn from it.[88] The Bible is also a place-centered book; she argues that it could not have been written in Babylon or Egypt because of the rich agricultural and geographical metaphors it uses. (Berry's notion that a "locally adapted land economy" implies a "locally adapted religion" may stem from this perspective.[89])

She suggests that the Bible "may open our eyes to recognize that land care is an area in which theologically informed moral discernment is needed."[90] She also

---

82 Ibid., 15.
83 Ibid., 21.
84 Ibid., 28.
85 Ibid., 38.
86 Ibid., 39.
87 Davis, *Scripture, Culture, and Agriculture,* 8.
88 Ibid., 29-30.
89 Ibid., xiii.
90 Ibid., 10.

notes that this is especially important due to the destruction of "the heartland" ("tellingly, now more often called the 'fly-over states'").[91]

Davis includes an analysis of what she calls "the biblical case for a local economy," suggesting that the biblical system is one whose "chief aim is the subsistence of local farming communities." [92] She states, "The ideal shared by many biblical writers... is that arable land is covenanted by God to the people Israel."[93] This land is not to be bought and sold but rather passed from generation to generation as an entitlement, what the Hebrew scriptures call *nahala.*[94] The shift in the United States away from family farms – only one percent of the population now makes a living from farming – also means that there has been a loss of the farming "vocation," and this is problematic because "from a biblical perspective, farming is the primary human vocation."[95]

Both Davis and Wirzba, then, provide contemporary Christianity with theological and biblical foundations for attending to agrarian concerns, and rural ministers have picked up on these themes. In *Sustaining Heart in the Heartland,* Miriam Brown and her coauthors from the interdenominational Rural Spirituality Team write, "It is important to the world that in the heartland people are working together on their part of the planet's critical questions of today. How do we love our neighbor? How do we care for the earth? How do we 'do' interdependence, sustainability dignity, and justice in a changing and interconnected world?"[96]

Writing in the early part of the first millennium, Rabbi 'Ahai ben Josiah made a passionate argument in favor of local food. He says:

> He who purchases grain in the market place, to what may he be likened? To an infant whose mother died: although he is taken from door to door to other wet nurses, he is not satisfied.
>
> He who buys bread in the marketplace, what is he like? He is as good as dead and buried.
>
> He who eats of his own is like an infant raised at his mother's breast.[97]

Though they might not express their beliefs in such strong terms, the linkage between land and people as understood in a religious sense, and as experienced in the heartland, makes a powerful case for the local foods movement in all of its forms.

---

[91] Ibid., 13.
[92] Ibid., 101.
[93] Ibid., 102.
[94] Ibid., 107.
[95] Ibid., 104.
[96] Brown et al., *Sustaining Heart in the Heartland,* ix.
[97] Davis, *Scripture, Culture, and Agriculture,* 21.

Chapter 15: Concluding Thoughts

Heather McIlvaine-Newsad

If we surrendered to earth's intelligence
We could rise up rooted like trees. – Rainer Maria Rilke

The modern US food system is extremely complex and polarized. On the one hand there is unparalleled concentration of corporate control over what can be planted, grown, bought and sold. On the other hand, the direct sales of foods to consumers through CSAs, farm-stands, farmers' markets, and other alternative outlets are growing (Hinrichs 2000). On February 19, 2013 the website Foodnavigator-usa.com released the following statement: "Food recalls in the US hit a two year high the fourth quarter…That comes to an average of six recalls a day, affecting about 18.4 million products." According to the report, 94 percent of the recalled foods fell within the Class 1 designation as items that can cause serious health consequences. Pinpointing the sources of foodborne illnesses become more difficult the more multifaceted our food system is. However, in the recent past our food system wasn't as complex as it is now.

As a little girl in the late 1960's and early 1970's, I remember spending days with my Grandma Mildred while my mom taught school and my dad was out in the fields. Some mornings, she would ask me, "Heather, what would you like for lunch?" Every now and then I would answer "fried chicken" and other times I would say "blackberry pie." What followed was never anything less than an adventure either to the chicken coop to find an appropriate candidate for the cast iron skillet; or over the hillside to find a bramble of ripe berries to form the basis of a fruity dessert. Sometimes there weren't any ripe blackberries or we weren't able to catch a chicken for lunch and one had to make do with what was in the cupboards or buttery. But that too was mostly local, mostly things my parents had canned or frozen or slaughtered from the previous summer. My point is, as a child I knew where my food came from. I understood that a chicken is a living, breathing animal that does not magically appear as frozen nuggets ready to heat up in the microwave for my lunch. If I got sick after a meal, if was much easier to find the source of illness through the simple process of identifying what I ate and where it came from. If I had bacon and eggs for breakfast, chances were that my Grandma Mildred knew where the eggs and bacon came from. This type of knowledge about one's food is so far removed from the reality of most people in America today. Unless you buy eggs that contain a stamp on them identifying the company that packaged and sold them to you, it is next to impossible to trace the sources of our food.

This book examines one small, but growing part of the food system in the United Sates - the local food movement in west central Illinois. The contributors in this book each approach his or her chapter from their particular academic disciplines and as active members of that movement. Some of the authors trace their interest in the movement based on childhood memories of what it felt like to dig in the dirt and grow something from seed to plant. Others approach their involvement in the local food movement as an issue of morality, which is by no means a new concept.

In 1826, Anthelem Brillat-Savarin published Physiologie du Gout, ou Meditations de Gastronomie *Transcendante* in which he wrote:

> "*Dis-moi ce que tu manges, je te dirai ce que tu es*."
> (Tell me what you eat and I will tell you what you are.)

In an essay titled *Concerning Spiritualism and Materialism*, 1863/4, Ludwig Andreas Feuerbach wrote:

> *"Der Mensch ist, was er ißt."*
> (Man is what he eats.)

Neither Brillat-Savarin or Feuerbach meant their quotations to be taken literally. They were implying that the food one eats has a bearing on what one's state of mind and health – in other words, one's morality. And like it or not what we choose to eat and from whom we choose to buy our food has moral implications ranging from environmental sustainability to social justice and animal welfare. The more mindfully we approach our food choices, the more aware we are of the ramifications – social, environmental, economic and other. And maybe if we are lucky we will make healthier food choices that will benefit our bodies as well.

So, I leave you with three rules that form the foundation of the local food movement and connect mind, body and spirit (Friedmann 2003):

1. Grow what is good for the earth.
2. Eat what is good for you.
3. Live in relationships that make the first two possible.

References

Anthelem Brillat-Savarin. 1842. *Physiologie du Gout, ou Meditations de Gastronomie Transcendante. Paris, France.*

Barrows, Anita and Joanna Maria Macy. 2005. Rilke's Book of Hours: Love Poems to God. Riverhead Trade.

Foodnavigator-usa.com
http://www.foodnavigator-usa.com/Business/Food-recalls-hit-2-year-high-in-fourth-

quarter?utm_source=copyright&utm_medium=OnSite&utm_campaign=copyright. Accessed February 25, 2013.

Friedmann, Harriet. 2003. Eating in the Garden of Gaia: Envisioning Polycultural Communities. *In* Fighting for the Farm: Rural American Transformed. Jane Adams, ed. 252-274. University of Pennsylvania Press.

Feuerbach, Ludwig Andreas 1863/4. Concerning Spiritualism and Materialism. Munich, Germany.

Hinrichs, Claire. 2000. The embeddedness of local food systems: Notes on two types of direct agricultural markets. *Journal of Rural Studies* 16:295-303.

CPSIA information can be obtained
at www.ICGtesting.com
Printed in the USA
FFOW01n1847151215
19297FF

9 781612 29196